SpringerBriefs in Computer Science

SpringerBriefs present concise summaries of cutting-edge research and practical applications across a wide spectrum of fields. Featuring compact volumes of 50 to 125 pages, the series covers a range of content from professional to academic.

Typical topics might include:

- A timely report of state-of-the art analytical techniques
- A bridge between new research results, as published in journal articles, and a contextual literature review
- A snapshot of a hot or emerging topic
- An in-depth case study or clinical example
- A presentation of core concepts that students must understand in order to make independent contributions

Briefs allow authors to present their ideas and readers to absorb them with minimal time investment. Briefs will be published as part of Springer's eBook collection, with millions of users worldwide. In addition, Briefs will be available for individual print and electronic purchase. Briefs are characterized by fast, global electronic dissemination, standard publishing contracts, easy-to-use manuscript preparation and formatting guidelines, and expedited production schedules. We aim for publication 8–12 weeks after acceptance. Both solicited and unsolicited manuscripts are considered for publication in this series.

**Indexing: This series is indexed in Scopus, Ei-Compendex, and zbMATH **

More information about this series at http://www.springer.com/series/10028

Huaqing Wu • Feng Lyu • Xuemin Shen

Mobile Edge Caching in Heterogeneous Vehicular Networks

 Springer

Huaqing Wu
University of Waterloo
Waterloo, ON, Canada

Xuemin Shen
Department of Electrical Computer
Engineering
University of Waterloo
Waterloo, ON, Canada

Feng Lyu
Computer Science
Central South University
Changsha, China

ISSN 2191-5768   ISSN 2191-5776 (electronic)
SpringerBriefs in Computer Science
ISBN 978-3-030-88877-0   ISBN 978-3-030-88878-7   (eBook)
https://doi.org/10.1007/978-3-030-88878-7

This Springer imprint is published by the registered company Springer Nature Switzerland AG
The registered company address is: Gewerbestrasse 11, 6330 Cham, Switzerland

Preface

The advanced technology of connected and automated vehicle (CAV) is expected to revolutionize the future transportation systems, which can enable vehicular information exchange and content delivery in real time. By utilizing cutting-edge technologies (e.g., advanced sensors, wireless communications and networking, and on-board processing), CAVs can be responsive to on-road emergencies rapidly and the road safety can be enhanced significantly. Furthermore, CAVs are also expected to support multifarious vehicular infotainment applications to improve the experience of both drivers and passengers. In particular, empowered by vehicle-to-everything (V2X) communications, safety-related and infotainment contents (such as road conditions, weather and traffic reports, news, video, music, and webpage) can be exchanged among vehicles or delivered to the vehicles from the infrastructure (such as roadside unit [RSU] and LTE base station [BS]) to provide comfortable driving experiences and content-rich multimedia services.

To facilitate smart vehicular services especially in the future driverless era, high-bandwidth content delivery and reliable accessibility of various applications are expected. However, current cellular networks cannot cope with the explosively growing mobile traffic demand to satisfy diversified content delivery services. To support tremendous vehicular content delivery, heterogeneous vehicular networks (HetVNets), which integrate the terrestrial networks with aerial networks formed by unmanned aerial vehicles (UAVs) and space networks constituting of low-Earth-orbit (LEO) satellites, can be utilized to provide seamless, robust, and reliable vehicular service provisioning. In addition, edge caching is an efficient solution to facilitate content delivery by caching popular files in HetVNet access points (APs) with one-hop content delivery from the caching-enabled APs to vehicles, which can mitigate the backhaul traffic and reduce the content delivery delay. However, it is challenging to achieve satisfying edge caching performance in HetVNets as various technical issues remain to be fully addressed. First, edge caching in HetVNet APs should jointly consider the differentiated file profiles and network characteristics (e.g., network coverage, network capacity, AP distribution) to fully unleash the potential of HetVNets for caching performance enhancement. Second, when UAVs are involved, the UAV trajectory should be jointly optimized with content placement and content delivery, which has not been well-addressed due to their complicated intercoupling relationships. Third, for caching-based HetVNet content delivery, different network segments should cooperatively serve vehicular content requests by leveraging heterogeneous network resources ingeniously, rendering the problem intractable. The scheme design of the cooperative content delivery and the corresponding resource allocation are crucial to content delivery performance yet challenging to be addressed. Fourth, as the vehicular network topology and service requests vary significantly with uncertainties, the decision-making system should be able to keep pace with the dynamic vehicular environments, posing real-time requirements to the optimization solutions.

In this monograph, we investigate mobile edge content caching and delivery in HetVNets to provide better service quality for vehicular users with resource utilization efficiency enhancement. In Chap. 1, we provide an overview of Het-VNets, including how it can support high-bandwidth content applications, technical challenges, and the research objective of this monograph. In Chap. 2, we review the state-of-the-art techniques for content delivery performance enhancement and organize a comprehensive survey related to the HetVNet-based content delivery and mobile edge caching-based techniques. In Chap. 3, a coding-based content caching scheme is designed for the terrestrial HetVNet with intermittent network connections, and a matching-based algorithm is proposed to optimize the content placement to minimize the average content delivery delay. In Chap. 4, UAVs with caching capabilities are leveraged to cache content files and serve the vehicular users. A joint caching and trajectory optimization problem is investigated to make decisions on content placement, content delivery, and UAVs' trajectories simultaneously. A deep learning-based algorithm is then proposed to enable real-time decision-making in the highly dynamic vehicular networks. In Chap. 5, a space–air–ground integrated vehicular network (SAGVN) is studied where LEO satellites, UAVs, and terrestrial networks cooperate to serve the vehicular users in content delivery. Considering the characteristics of different network segments, a cooperative content delivery scheme is designed to jointly optimize the user association, bandwidth allocation, and content delivery ratio. At last, we conclude this monograph and provide potential future research directions in Chap. 6. The systematic principle in this monograph provides valuable insights on the mobile edge caching scheme design and efficient exploitation of the heterogeneous network resources to fully unleash their differential merits in vehicular networks.

We would like to thank Dr. Jiayin Chen, Dr. Weisen Shi, and Conghao Zhou from the Broadband Communications Research (BBCR) group at the University of Waterloo, Prof. Ning Zhang from the University of Windsor, Prof. Li Wang from the Beijing University of Posts and Telecommunications, Prof. Haibo Zhou from the Nanjing University, Prof. Nan Cheng from the Xidian University, and Prof. Wenchao Xu from the Hong Kong Polytechnic University for their contributions in the presented research works. We also would like to thank all the members of BBCR group for the valuable suggestions and comments. In addition, Huaqing Wu would like to personally thank her husband, Mr. Yuhui Lin, for his heartfelt support and continuous encouragement throughout her career. Special thanks also go to the staff at Springer Nature: Mary E. James and Shabib Shaikh, for their help throughout the publication preparation process.

Waterloo, ON, Canada   Huaqing Wu

Changsha, China   Feng Lyu

Waterloo, ON, Canada   Xuemin Shen

Contents

Acronyms

CAV	Connected and automated vehicle
HetVNet	Heterogeneous vehicular network
RSU	Roadside unit
BS	Base station
UAV	Unmanned aerial vehicle
V2X	Vehicle-to-everything
V2V	Vehicle-to-vehicle
V2I	Vehicle-to-infrastructure
VN	Vehicular network
SAG	Space–air–ground
DSRC	Dedicated short-range communication
QoS	Quality of service
SAGVN	Space–air–ground integrated vehicular network
RAT	Radio access technology
TVWS	TV white space
HAP	High altitude platform
LAP	Low altitude platform
LoS	Line-of-sight
GEO	Geostationary Earth orbit
MEO	Medium Earth orbit
LEO	Low Earth orbit
AP	Access point
CBS	Cellular base station
ILP	Integer linear programming
AGVN	Aerial–ground vehicular network
JCTO	Joint caching and trajectory optimization
DSL	Deep supervised learning
CBTL	Clustering-based two-layered
CNN	Convolutional neural network
ABC	User association, bandwidth allocation, and content delivery ratio
LMA-ABC	Load- and mobility-aware ABC

DRL	Deep reinforcement learning
3GPP	3rd generation partnership project
CR	Cognitive radio
SDN	Software-defined networking
NFV	Network function virtualization
AI	Artificial intelligence
CDN	Content delivery network
QoE	Quality of experience
SE	Spectral efficiency
PRAI	Partial repeat after interruption
SA	Student admission
GS	Gale–Shapley
PSO	Particle swarm optimization
IoV	Internet of vehicle
U2V	UAV-to-vehicle
RCSP	Resource constrained shortest path
SNR	Signal-to-noise ratio
S2V	Satellite-to-vehicle
B2V	BS-to-vehicle

Chapter 1
Introduction

Abstract The advanced technology of connected and automated vehicles (CAVs) enables vehicles to interact with their internal and external environments to improve road safety, transportation efficiency, and the experience of both drivers and passengers. To empower smart vehicular services especially in the future driverless era, the CAV networks are expected to support high-bandwidth content delivery and reliable accessibility of multifarious applications. However, the limited radio spectrum resources, the inflexibility in accommodating dynamic traffic demands, and geographically constrained fixed infrastructure deployment of current terrestrial networks pose great challenges in ensuring ubiquitous, flexible, and reliable network connectivity. To address these challenges in a cost-effective way, heterogeneous vehicular networks (HetVNets) that integrate terrestrial networks with non-terrestrial networks can be leveraged to boost network capacity, enhance system robustness, and provide ubiquitous 3D wireless coverage. Furthermore, mobile edge caching technologies can be utilized in HetVNets to further mitigate backhaul traffic burden and reduce vehicular content delivery delay. In this chapter, we first provide an overview of the vehicular content delivery networks and then elaborate the mobile edge caching-assisted HetVNets with differentiated network characteristics. Finally, we present the key research problems investigated in this monograph.

1.1 Overview of Vehicular Content Delivery Networks

With the tremendous technological development in advanced sensors, on-board processing, and wireless communications and networking, connected and automated vehicles (CAVs) are expected to perform essential roles in diversified fields of human society. As predicted by International Data Corporation, the number of CAVs will continue to surge over the next several years, increasing from 31.4 million units in 2019 to 54.2 million units (more than 50% of all vehicles produced) in 2024 [1]. Information exchange for CAVs can be realized via both intra-vehicle communications and inter-vehicle/vehicle-to-infrastructure communications [2]. Intra-vehicle communications happen within a vehicle, such as among different on-board sensors and systems. Via inter-vehicle/vehicle-to-infrastructure communications, a vehicle

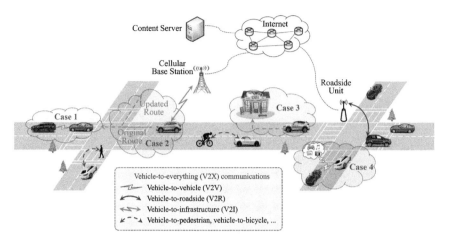

Fig. 1.1 An overview of vehicular content delivery networks

can communicate with other entities such as other vehicles, roadside units (RSUs), and base stations (BSs), collectively referred to as vehicle-to-everything (V2X) communications, as shown in Fig. 1.1. With V2X communications, ubiquitous information exchange and content delivery can be enabled to support multifarious CAV applications. In particular, content delivery in vehicular networks (VNs) has the following typical application scenarios [3]:

1. **Safety Information Delivery:** Road safety is of the utmost importance for connected vehicles [4]. To reduce the frequency and severity of vehicle collisions, vehicles need to monitor the environmental data and collect vehicle state data to evaluate vehicles' safety status. Vehicles' safety-related information includes the available energy (e.g., fuel and electric), moving direction and speed, distances among vehicles, etc. By analyzing the collected safety-related information, the system can deliver alarm information when necessary and perform accident forewarning, shown as **Case 1** in Fig. 1.1.

2. **Traffic Efficiency Information Delivery:** To facilitate traffic management and transportation efficiency, useful information (including vehicles' locations, road conditions, and congestion situations) should be exchanged for system coordination and route planning. For example, on-board map applications can automatically reroute according to the received traffic information to avoid the congestion area and enhance traffic efficiency, shown as **Case 2** in Fig. 1.1.

3. **Infotainment Content Delivery:** To improve drivers' and passengers' traveling experiences, the infotainment content (such as news, music, and videos) can be delivered and shared to provide informative or entertaining services. For example, service providers can collect vehicles' real-time locations and users' tastes and health conditions to recommend a restaurant and deliver the related information such as public evaluation and open hours, shown as **Case 3** in Fig. 1.1. Furthermore, vehicles in proximity may request similar location-based

content such as traffic conditions and map data. In such cases, data transmission can be accomplished with direct delivery among adjacent vehicles, shown as **Case 4** in Fig. 1.1.

In this monograph, we mainly investigate the delivery of infotainment content. Notice that content delivery in VNs is different from that in conventional mobile networks since it is highly dependent on vehicles' mobility patterns, road conditions, and user behaviors [5]. Considering the massive amount of vehicular data, dynamic vehicular network conditions, and differentiated service requirements, there exist various technical challenges for content delivery in VNs:

1. *Access network congestion:* With the increasing number of vehicles on road and the proliferation of multifarious vehicular applications, the vehicular data requirement is soaring at a tremendous pace. According to the mobile data forecast from Cisco, the mobile data traffic will grow at a compound annual growth rate (CAGR) of 46%, increasing sevenfold from 2017 to 2022 [6]. Furthermore, connected cars will be the fastest-growing industry segment with a CAGR of 28%. The Federal Communications Commission has allocated 75 MHz bandwidth at the 5.9 GHz spectrum band to dedicated short-range communications (DSRCs) for vehicular communications. However, DSRC mainly focuses on enabling vehicular safety applications, i.e., supporting rapid short message exchange, and the limited spectrum resource is insufficient to satisfy the quality of service (QoS) of the bandwidth-intensive infotainment applications. In addition, the cellular network capacity is not able to grow at a comparable pace to support the enormous data traffic due to the scarce spectrum resources and high cost of the infrastructure upgrade. Thus, exploring alternative networks to support infotainment content delivery is imperative to alleviate the access network congestion problem.

2. *Backhaul network congestion:* In addition to access network congestion, the transmission of massive vehicular data also increases the probability of blocking the backhaul networks, especially for wireless backhaul networks that have limited capacities and are easily congested. Although wired backhaul networks can provide a higher transmission data rate and have more spectrum resources, it is costly and sometimes difficult to deploy especially in some remote or mountainous areas.

3. *On-demand service provisioning:* Network conditions (e.g., traffic density and request distribution) in VNs are highly dynamic in both temporal (e.g., peak hours or midnight) and spatial (e.g., urban or rural areas) domains. To guarantee uniform service coverage, the infrastructure needs to be densely deployed, which significantly increases the deployment cost. Therefore, the efficient and cost-effective infrastructure deployment in VNs to provide on-demand content delivery services in different demanding areas is a challenging problem.

4. *Global connectivity and reliability:* Global network connectivity, which is essential for ubiquitous vehicular service provisioning, can hardly be achieved by relying only on the current terrestrial networks due to the geographically constrained infrastructure deployment. For example, it is cost-ineffective or even

impossible to deploy infrastructure in sparsely populated or remote mountainous areas. Furthermore, how to guarantee reliable and uninterrupted network connections for vehicular content delivery regardless of the potential infrastructure outage (e.g., caused by natural disasters) is another critical yet challenging topic.

In summary, VNs can improve road safety and provide better travel experiences for drivers and passengers by enabling safety-related and infotainment content delivery. However, the massive traffic demands, the inherent characteristics of VNs, and the diversified vehicular service requirements make it challenging to design an efficient, flexible, and cost-effective content delivery system. To address the above-mentioned challenges, in this monograph, we investigate the mobile edge caching-assisted content delivery in HetVNets to improve the service quality with enhanced resource utilization efficiency.

1.2 Mobile Edge Caching-Assisted Content Delivery in HetVNets

To support tremendous CAV content delivery, HetVNets are considered to expand the breadth and depth of communication coverage by utilizing multiple revolutionary networking techniques. Specifically, aerial networks based on unmanned aerial vehicles (UAVs) and space networks consisting of satellites can be involved in HetVNets to assist vehicular content delivery, thus achieving a space–air–ground integrated vehicular network (SAGVN). With the integration of terrestrial, aerial, and space networks, the SAGVN can exploit the complementary advantages of different network segments to provide globally seamless, reliable, flexible, and cost-effective network access [7–9].

Despite the benefits and potentials brought by the SAGVN, the backhaul transmission of the vehicular traffic data still faces some technical challenges: (1) for terrestrial HetVNets, although the wireless cellular traffic burden can be relieved by utilizing other radio access technologies (RATs) like TV white space (TVWS) and Wi-Fi, the backhaul networks that support all vehicular traffic data still suffer a high congestion probability, (2) for UAV-assisted network access, the wireless backhaul links are generally slow and unreliable, and (3) for satellite-based content delivery, an unacceptable delivery delay may occur if the satellite goes through backhaul links to retrieve the requested content due to the long propagation delay. To address these issues, mobile edge caching technologies can be utilized to cache content files closer to the end users to alleviate backhaul congestion, reduce energy consumption, and decrease content retrieving delay.

In this section, we first introduce the heterogeneous SAGVN where different network segments have differentiated network characteristics. Then, mobile edge caching-assisted content delivery in the SAGVN will be presented.

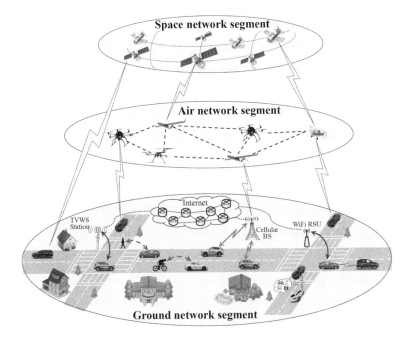

Fig. 1.2 An illustration for space–air–ground integrated vehicular networks

1.2.1   Overview of the Heterogeneous SAGVN

As shown in Fig. 1.2, the SAGVN compromises three main network segments: ground networks, aerial networks, and space networks. The integration of these network segments has attracted increasing attention from both academia and industry. In this part, we first summarize the communication characteristics in different network segments. Then, advantages of the integrated SAGVN will be introduced.

Ground Networks As the main solution to provide wireless network coverage in most scenarios, ground networks consist of heterogeneous terrestrial communication systems such as cellular networks, Wi-Fi, TVWS, and millimeter-wave communication networks [10]. Ground networks can be characterized by the following features:

1. *Ultra-dense small cells:* As the key component in ground networks, cellular networks have evolved rapidly from the first generation (1G) to the fifth generation (5G) wireless networks. With the network development, the cell size becomes smaller with increasing BS density to enhance spectrum efficiency and network throughput. However, the ultra-dense deployment of small cells also leads to a high construction and maintenance cost.

2. *High-speed fiber links:* Ground nodes in the backbone network are interconnected via fiber optic links, which can provide low-cost high-throughput data transmissions.
3. *High-performance computing and caching:* The data centers and servers in ground networks generally have powerful computing capabilities and massive caching resources, which can ensure efficient network operation and management to provide multifarious services.
4. *Fixed and limited coverage:* Basically, the ground infrastructure is fixedly deployed. Therefore, ground network coverage is limited and geographically constrained, especially in sparsely populated or remote mountainous areas.

Aerial Networks Aerial networks are formed by UAVs at different altitudes, including the high altitude platform (HAP) which generally operates at an altitude of 17–22 km and the low altitude platform (LAP) which typically works at an altitude of no more than several kilometers. Comparing LAP and HAP networks, HAP networks can provide a wider coverage with a longer endurance, while LAP networks are more flexibly deployed and configured with better short-range communication performance. In this monograph, we focus on LAP-UAVs to assist terrestrial networks in content delivery.[1] With the inherent characteristics of flexibility and controllability, UAVs have been considered as an indispensable component and a promising technique in the next generation networks due to the following advantages:

1. *Line-of-sight (LoS) communication links:* Compared to terrestrial communication links, there exist fewer obstacles between UAVs and ground vehicles due to the high UAV operation altitude. With a higher probability of LoS connections, UAV communication links are generally more reliable with a better communication performance.
2. *Flexible deployment:* Different from the fixed deployment of terrestrial BSs, UAVs can be dynamically deployed to adapt to the spatially and temporally varying ground traffic. Since UAVs can be dispatched to different areas on demand, it is more cost-effective than deploying static BSs to guarantee QoS.
3. *Fully controlled mobility:* Leveraging the agility and full controllability, UAVs' positions can be adjusted and optimized in real time to improve communication link quality and respond to potential emergency situations.

Despite the advantages of UAVs, there exist technical issues affecting the UAV communication performance. First of all, UAVs are battery-powered and energy-constrained, and the propulsion and directional adjustment consume most of the energy. Thus, energy efficiency in UAV communications is critical to guarantee long-term network access without service interruption. In addition, extreme weather conditions can also affect the UAV operation and should be considered in UAV communications.

---

[1] In the rest of this monograph, "UAVs" are used to refer to "LAP-UAVs" for simplicity.

Space Networks Space networks are composed of satellites in different orbits, i.e., geostationary Earth orbit (GEO), medium Earth orbit (MEO), and low Earth orbit (LEO) satellites. Although GEO and MEO satellites can provide a large coverage with a low relative velocity between satellites and terrestrial users, the excessive propagation delay inhibits their applicability in most time-sensitive vehicular applications. In recent years, LEO satellite communications have attracted significant attention and are deemed as promising solutions to be incorporated in future network architectures due to the following advantages:

1. *Globally seamless coverage:* Benefiting from global availability, LEO networks have the potential to capacitate worldwide seamless service coverage in a cost-effective way, especially for users dispersed over wide geographical areas or in inaccessible areas.
2. *Low delay for long-distance communications:* Due to the low orbit altitude (300~1500 km) compared to GEO (35,786 km) and MEO (7000~25,000 km) satellites, the one-way LEO communication propagation delay is less than 14 ms. Users can experience a round-trip delay of no more than 50 ms, which is comparable to that of terrestrial links [12]. For long-distance communications, satellite communications can achieve even lower delay than terrestrial links due to the small number of communication hops.
3. *Enhanced communication efficiency:* The inherent broadcast/multicast nature of satellite communications enables group-based transmissions to enhance communication efficiency. For example, satellite communications can support CAV software update simultaneously for millions of vehicles with negligible communication cost.
4. *High network reliability and robustness:* Due to the invulnerability to natural disasters, space networks can support ultra-reliable and robust service provisioning.

However, similar to UAVs, the limited energy capacities of LEO satellites also constrain the service duration and affect service functionalities including sensing, transmission, and processing. In addition, the high speed of LEO satellites (up to 28,000 km/h relative to the Earth's surface) results in highly dynamic network topologies, intermittent network connections, and frequent handover.

Integration of Space–Air–Ground Networks As mentioned above, different network segments have their pros and cons in service provisioning, such as in terms of coverage, transmission delay, throughput, and reliability, as summarized in Table 1.1. With effective internetworking, the complementary advantages of different network segments can be leveraged to enhance vehicular content delivery service qualities. For instance, satellite communications can supplement terrestrial networks for service provisioning in remote or sparsely populated areas; meanwhile, the complementary properties of satellite links (wide coverage) and the fiber optic backbone (high data rate) can be considered as alternative backbone technologies to terrestrial wireless backhaul to alleviate the long-distance multi-hop backhaul. UAVs can provide flexible and reliable connectivity for vehicles in congested areas

Table 1.1 Characteristics of terrestrial, aerial, and space network segments

	Terrestrial networks	Aerial networks	Space networks
System deployment	Fixed deployment before use	Flexible deployment on demands	System configuration required with a long lead time
Mobility	Static	30–460 km/h [11]	28,000 km/h at orbits of 200 km (reduce with increased altitude)
Propagation delay	Low	Low	Less than 14 ms [12]
Advantages	High throughput and powerful computing and caching capabilities	Low cost, flexible movement, high LoS link probability	Large coverage, ultra-reliability, broadcast/multicast capabilities
Disadvantages	Limited coverage, inflexible deployment, vulnerable to disaster	Energy-constrained, limited capacity, high mobility	Limited capacity, high mobility, non-negligible propagation delay

with dynamic traffic demands to relieve the terrestrial network burden and to boost service capacity. In addition, satellites/UAVs with remote sensing technologies can provide large-scale monitoring data to assist terrestrial networks for efficient resource management and planning decisions. Therefore, the SAGVN can ensure seamless, robust, and reliable vehicular service provisioning in a cost-effective way.

1.2.2 Mobile Edge Caching-Assisted Content Delivery

Despite the tremendous potentials of the SAGVN, the backhaul limitations (e.g., limited capacity and long propagation delay) may degrade the performance of vehicular content delivery. Stemming from the observations in [13], a large portion of mobile multimedia traffic can be attributed to duplicated downloads of a small fraction of popular content files. Furthermore, duplicated content requests can be intensified in small regions with certain events (e.g., concerts or sports games) where people have common interests in hot content. In VNs, location-based applications boost the repetitive download of location-oriented data (e.g., real-time traffic reports, high definition maps, and so forth). To improve the computing and storage resource utilization of current network infrastructure and modern vehicles, it is promising to cache the popular content closer to end users. By enabling direct content delivery from the caching-enabled access points (APs), e.g., Wi-Fi RSUs, TVWS stations, cellular base stations (CBSs), UAVs, and LEO satellites, to the vehicles, content caching on network edge infrastructure can significantly off-load backhaul traffic [14]. Besides, by avoiding the repeated backhaul transmission, the content delivery delay can be significantly reduced, which is essential in VNs to facilitate efficient content delivery with rapidly changing network topology.

When designing the mobile edge caching strategies, there are two fundamental building blocks: *content placement (or content caching[2])* and *content delivery*. Content placement mainly concentrates on determining content files to be cached and the selection of appropriate caching nodes to store content files. Basically, the design of content placement schemes is related to the management of caching resources with the objective of serving more users' content requests with better content delivery performance. Content delivery, on the other hand, focuses on the dissemination of cached content from caching nodes to requesting users. Content delivery is mainly about the management of communication resources including the design of routing and forwarding policies, spectrum allocation schemes, and power control strategies.

In the literature, there have been some existing works on the vehicular content delivery in the edge caching-assisted HetVNet to achieve better delivery performance. In the following, we briefly list the main research issues in mobile edge content caching and delivery, as summarized in Table 1.2:

1. *Where to cache:* In the SAGVN, the potential caching nodes include vehicles and/or the edge infrastructure (e.g., CBSs, Wi-Fi RSUs, TVWS stations, UAVs, and LEO satellites) in different network segments. The selection of caching nodes is affected by multiple factors including but not limited to the energy and storage capacity constraints of caching nodes, the willingness of caching nodes to share their resources, and the contact characteristics between the requesting vehicles and the caching nodes.
2. *What to cache:* The problem of what to cache is largely determined by content file sizes, the storage capacity of caching nodes, and content popularity distributions. Specifically, the content popularity is generally assumed to follow static models (e.g., Zipf model [15]) or dynamic models (e.g., shot noise model (SNM) [16]). In recent years, machine learning (ML)-based content popularity prediction methods have also attracted significant attention [17, 18].
3. *How to cache:* Based on the knowledge of content popularity and network information (such as network topology and user mobility), various caching policies and algorithms have been proposed to enhance caching efficiency. Conventional caching policies such as the least recently used (LRU) and least frequently used (LFU) policies have been widely utilized in existing works [19]. User preference profile and vehicle mobility information can also be utilized for caching scheme design to enhance the caching performance [20, 21].
4. *Where to obtain the content:* With appropriate content files cached in caching nodes, the association between the caching nodes and the content requesters should also be investigated to optimize the content delivery performance.

---

[2] In this monograph, we use the terms content caching and content placement interchangeably.

Table 1.2 Summary on content placement and delivery

	Research issues	Description	Required information
Content placement	Where to cache	• Caching nodes: vehicles/BSs/UAVs/LEO satellites • Caching node placement • Caching node selection	• (Expected) network topology • Content request pattern • User mobility • Content size • Content popularity • Caching storage capacity
	What to cache	• Reusable information (e.g., multimedia data) can be cached, while interactive applications, voice calls, or control signals cannot be cached • Determining which content files to cache	
	How to cache	• Deciding files to be cached in different caching nodes • Caching storage allocation	
Content Delivery	Where to obtain the content	• Association between caching nodes and content requesters	• (Expected) network topology • Content request pattern • User mobility and content size • Network condition: network capacity, link quality, congestion condition, etc.
	How to deliver the content	• Routing path selection • Communication resource (e.g., power and bandwidth) allocation	

Basically, the caching node–requester association problem can be formulated as an integer programming problem and solved by matching-based algorithms [14, 15].

5. *How to deliver the content:* The content delivery scheme involves the problem of selecting the optimal delivery path and the corresponding resource allocation. In [22], the optimal content delivery is investigated in caching-enabled HetVNets with stochastic content multicast scheduling to satisfy dynamic user demands. The signal transmission has also been optimized for content delivery to minimize the energy consumption in caching-assisted HetVNets [23].

1.3   Technical Challenges

In spite of the initial research works mentioned above, efficient mobile edge content caching and delivery in the SAGVN are still not sufficiently studied. To be specific, the design and implementation of the edge caching-assisted SAGVN still face some essential challenges:

1. *Supporting high mobility:* In the SAGVN, there are various types of mobility introduced by vehicular users, UAVs, and LEO satellites. As a result, the contact duration between a vehicular user and an AP (e.g., CBS, Wi-Fi RSU, UAV, LEO satellite) in the SAGVN is limited. Such limited contact duration may be insufficient for content delivery, especially for large-size content files. Therefore, caching the whole content files in APs is inefficient since the complete downloading of one file requires multiple encounters with APs. The edge caching scheme design considering the user mobility and content file size should be investigated to improve caching efficiency.

2. *Heterogeneous network characteristics:* As introduced in Sect. 1.2.1, different network segments in the SAGVN have their specific pros and cons in terms of coverage, throughput, propagation delay, etc. The unprecedented heterogeneity in the SAGVN renders the traditional caching schemes inadequate. It is critical to customize the content caching and delivery policies by considering heterogeneous network characteristics to fully unleash their differential merits.

3. *Coupled UAV trajectory design and resource allocation:* Note that UAVs' ability to adapt to terrestrial traffic variations is capacitated by the fully controllable mobility. Therefore, the UAV trajectory design problem, which optimizes UAVs' flying traces to serve vehicular users, is critical for UAV-assisted service provisioning. On the other hand, the allocation of both caching and communication resources should also be optimized to improve the UAV content delivery performance in vehicular networks. Therefore, content caching, UAV trajectory, and content delivery are tightly coupled and should be jointly investigated, which has not been well addressed.

4. *Energy constraints:* Different from the terrestrial infrastructure that has a stable power supply, UAVs/satellites are powered by batteries and/or solar energy. The limited energy capacity of UAVs/satellites constrains the service endurance and further affects service functionalities such as sensing, transmission, and processing. In addition, service-unrelated energy consumption for UAVs (due to propulsion and direction adjustment) and satellites (due to intense radiation and space-variant temperature) further deteriorates the service duration. Thus, vehicular users in the SAGVN can potentially suffer intermittent connections and service interruptions due to UAV/satellite energy depletion. Improving energy efficiency to prolong the service duration of UAVs/satellites is crucial for persistent vehicular content delivery service provisioning.

Besides the technical challenges, many countries and international organizations have specified some regulatory rules in terms of the usage of UAVs [24] and satellites [25]. To ensure the legal use of UAVs, all UAVs should be controlled by UAV operators or governments rather than individual users. According to the International Telecommunication Union (ITU) regulations, the emission energy from satellites should be regulated to prevent harmful interference to the incumbent terrestrial communication systems. International regulations and standards are imperative to ensure the proper operation of the SAGVN system. Nevertheless, in this monograph, we focus on technical challenges faced by mobile edge content caching and delivery in the SAGVN.

1.4 Aim of the Monograph

In this monograph, we investigate the mobile edge content caching and delivery schemes in the caching-enabled SAGVN to tackle the above-mentioned challenges. The objective is to develop efficient caching policies and design cooperative content delivery schemes to increase the network capability, improve resource utilization efficiency, and enhance service quality. Firstly, we review the state-of-the-art content caching and delivery schemes in the HetVNet and provide a comprehensive survey to state our technical motivations. Then, we propose and design content caching and delivery schemes in different HetVNet scenarios to improve the content delivery performance.

In Chap. 3, we first investigate the content caching scheme design in terrestrial HetVNets with fixed APs, where CBSs, Wi-Fi RSUs, and TVWS stations can cache popular content files to support vehicular content delivery [26]. Considering the limited network coverage and high vehicle mobility, an on–off model with service interruptions is established to characterize the intermittent vehicular network connections. To resist the impact of unstable network connections, content coding is utilized to enhance caching efficiency. By jointly considering file characteristics and network conditions, we formulate the content placement problem as an integer linear programming (ILP) problem to minimize the average content retrieving delay. Adopting the idea of the student admission model, we then transform the ILP problem into a many-to-one matching problem between content files and HetVNet APs and propose a stable matching-based caching scheme to solve it. Simulation results demonstrate that the proposed scheme can achieve near-optimal performances in terms of delivery delay and offloading ratio with a low complexity.

Focusing on the aerial–ground vehicular network (AGVN), UAV-aided caching is considered to assist vehicular content delivery in Chap. 4. To maximize the overall network throughput, a joint caching and trajectory optimization (JCTO) problem is investigated to jointly optimize content caching, content delivery, and UAV trajec- tories [27]. Considering the intercoupled optimization variables and limited UAV on-board energy, the JCTO problem is intractable directly and timely. To enable real-time decision-making in highly dynamic vehicular networks, we propose a deep supervised learning (DSL) scheme to solve the JCTO problem. Specifically, a clustering-based two-layered (CBTL) algorithm is first designed to solve the

JCTO problem offline. With a given content caching policy, we design a time-based graph decomposition method to jointly optimize content delivery and UAV trajectory, with which the particle swarm optimization algorithm is then leveraged to further optimize the content caching. We then design a DSL architecture of the convolutional neural network (CNN) to make online decisions. The network density and content request distributions with spatial–temporal variations are labeled as channeled images and input to the CNN-based model, and the results obtained from the CBTL algorithm are labeled as model outputs. The CNN-based model can be trained to intelligently learn the mapping function between the input network information and output decisions and make real-time inferences. Extensive trace-driven experiments are conducted to demonstrate the efficiency of CBTL in solving the JCTO problem and the superior learning performance with the CNN-based model.

In Chap. 5, we focus on the vehicular content delivery in the SAGVN [28, 29]. To improve the vehicular content delivery performance, we investigate caching-assisted cooperative content delivery to minimize the overall content retrieving delay. In particular, vehicular content requests can be cooperatively served by multiple APs in space, aerial, and terrestrial networks. A joint optimization problem of vehicle-to-AP *a*ssociation, *b*andwidth allocation, and *c*ontent delivery ratio, referred to as the *ABC* problem, is then formulated. To address the tightly coupled optimization variables, we propose a load- and mobility-aware *ABC (LMA-ABC)* scheme to solve the joint optimization problem as follows. We first decompose the *ABC* problem to optimize the content delivery ratio by considering the distinct characteristics of different network segments. Then the impact of bandwidth allocation on the achievable delay performance is analyzed, and an effect of diminishing delay performance gain is revealed. Based on the analysis results, the *LMA-ABC* scheme is designed with the consideration of user fairness, load balancing, and vehicle mobility. Simulation results demonstrate that the proposed *LMA-ABC* scheme can significantly reduce the cooperative content delivery delay compared to the benchmark schemes.

Finally, we conclude this monograph and discuss future research directions in Chap. 6, including service-oriented multi-dimensional resource orchestration, software-defined networking (SDN)-based control architecture in the SAGVN, and artificial intelligence (AI)-based network management.

References

1. International Data Corporation (IDC), A new IDC forecast shows how vehicles will gradually incorporate the technologies that lead to autonomy. https://www.idc.com/getdoc.jsp?containerId=prUS46887020&utm_medium=rss_feed&utm_source=Alert&utm_campaign=rss_syndication. Accessed July 2021
2. J. Wang, J. Liu, N. Kato, Networking and communications in autonomous driving: a survey. IEEE Commun. Surv. Tut. **21**(2), 1243–1274 (2018)

3. Z. MacHardy, A. Khan, K. Obana, S. Iwashina, V2X access technologies: regulation, research, and remaining challenges. IEEE Commun. Surv. Tut. **20**(3), 1858–1877 (2018)
4. F. Lyu, N. Cheng, H. Zhu, H. Zhou, W. Xu, M. Li, X. Shen, Towards rear-end collision avoidance: adaptive beaconing for connected vehicles. IEEE Trans. Intell. Transp. Syst. 22(2), 1248–1263 (2021)
5. C. Xu, Z. Zhou, Vehicular content delivery: A big data perspective. IEEE Wireless Commun. 25(1), 90–97 (2018)
6. Cisco, Cisco visual networking index: global mobile data traffic forecast update, 2017–2022. https://s3.amazonaws.com/media.mediapost.com/uploads/CiscoForecast.pdf. Accessed July 2021
7. N. Zhang, S. Zhang, P. Yang, O. Alhussein, W. Zhuang, X. Shen, Software defined space-air-ground integrated vehicular networks: challenges and solutions. IEEE Commun. Mag. 55(7), 101–109 (2017)
8. T. Hong, W. Zhao, R. Liu, M. Kadoch, Space-air-ground IoT network and related key technologies. IEEE Wireless Commun. 27(2), 96–104 (2020)
9. F. Lyu, F. Wu, Y. Zhang, J. Xin, X. Zhu, Virtualized and micro services provisioning in space-air-ground integrated networks. IEEE Wireless Commun. 27(6), 68–74 (2020)
10. W. Wu, N. Cheng, N. Zhang, P. Yang, W. Zhuang, X. Shen, Fast mmwave beam alignment via correlated bandit learning. IEEE Trans. Wireless Commun. **18**(12), 5894–5908 (2019)
11. H. Nawaz, H.M. Ali, A.A. Laghari, UAV communication networks issues: a review. Arch. Comput. Methods Eng. **28**, 1349–1369 (2021)
12. B. Di, L. Song, Y. Li, H.V. Poor, Ultra-dense LEO: integration of satellite access networks into 5G and beyond. IEEE Wireless Commun. **26**(2), 62–69 (2019)
13. X. Wang, M. Chen, T. Taleb, A. Ksentini, V.C.M. Leung, Cache in the air: exploiting content caching and delivery techniques for 5G systems. IEEE Commun. Mag. **52**(2), 131–139 (2014)
14. L. Wang, H. Wu, Y. Ding, W. Chen, H.V. Poor, Hypergraph based wireless distributed storage optimization for cellular D2D underlays. IEEE J. Sel. Areas Commun. **34**(10), 2650–2666 (2016)
15. L. Wang, H. Wu, Z. Han, P. Zhang, H.V. Poor, Multi-hop cooperative caching in social IoT using matching theory. IEEE Trans. Wireless Commun. **17**(4), 2127–2145 (2018)
16. S. Traverso, M. Ahmed, M. Garetto, P. Giaccone, E. Leonardi, S. Niccolini, Temporal locality in today's content caching: why it matters and how to model it. ACM SIGCOMM Comput. Commun. Rev. **43**(5), 5–12 (2013)
17. N. Garg, M. Sellathurai, V. Bhatia, B. Bharath, T. Ratnarajah, Online content popularity prediction and learning in wireless edge caching. IEEE Trans. Commun. **68**(2), 1087–1100 (2020)
18. W.-X. Liu, J. Zhang, Z.-W. Liang, L.-X. Peng, J. Cai, Content popularity prediction and caching for ICN: a deep learning approach with SDN. IEEE Access **6**, 5075–5089 (2017)
19. A. Ioannou, S. Weber, A survey of caching policies and forwarding mechanisms in information-centric networking. IEEE Commun. Surv. Tut. **18**(4), 2847–2886 (2016)
20. H. Ahlehagh, S. Dey, Video-aware scheduling and caching in the radio access network. IEEE/ACM Trans. Netw. **22**(5), 1444–1462 (2014)
21. R. Wang, X. Peng, J. Zhang, K.B. Letaief, Mobility-aware caching for content-centric wireless networks: modeling and methodology. IEEE Commun. Mag. **54**(8), 77–83 (2016)
22. B. Zhou, Y. Cui, M. Tao, Stochastic content-centric multicast scheduling for cache-enabled heterogeneous cellular networks. IEEE Trans. Wireless Commun. **15**(9), 6284–6297 (2016)
23. F. Guo, H. Zhang, X. Li, H. Ji, V.C.M. Leung, Joint optimization of caching and association in energy-harvesting-powered small-cell networks. IEEE Trans. Veh. Technol. **67**(7), 6469–6480 (2018)
24. L. Gupta, R. Jain, G. Vaszkun, Survey of important issues in UAV communication networks. IEEE Commun. Surv. Tut. **18**(2), 1123–1152 (2015)
25. T. Xia, M.M. Wang, X. You, Satellite machine-type communication for maritime internet of things: an interference perspective. IEEE Access **7**, 76404–76415 (2019)

26. H. Wu, J. Chen, W. Xu, N. Cheng, W. Shi, L. Wang, X. Shen, Delay-minimized edge caching in heterogeneous vehicular networks: a matching-based approach. IEEE Trans. Wireless Commun. **19**(10), 6409–6424 (2020)
27. H. Wu, F. Lyu, C. Zhou, J. Chen, L. Wang, X. Shen, Optimal UAV caching and trajectory in aerial-assisted vehicular networks: a learning-based approach. IEEE J. Sel. Areas Commun. **38**(12), 2783–2797 (2020)
28. H. Wu, J. Chen, C. Zhou, F. Lyu, N. Zhang, L. Wang, X. Shen, Load- and mobility-aware cooperative content delivery in sag integrated vehicular networks, in *Proceedings of IEEE International Conference on Communications (ICC'21), Montreal* (2021)
29. H. Wu, J. Chen, C. Zhou, W. Shi, N. Cheng, W. Xu, W. Zhuang, X. Shen, Resource management in space-air-ground integrated vehicular networks: SDN control and AI algorithm design. IEEE Wireless Commun. **27**(6), 52–60 (2020)

Chapter 2
Techniques for Content Delivery Performance Enhancement

Abstract As the performance of vehicular content delivery can be enhanced by carefully designing content delivery schemes and content caching schemes, in this chapter, we provide a comprehensive survey of techniques for vehicular content delivery performance enhancement. Particularly, we present the state-of-the-art literature review in two sections. First, we present existing works on the heterogeneous vehicular networking (HetVNet) techniques, where multiple alternative networking techniques are utilized to off-load the access networks. Specifically, Wi-Fi and TV white space-based techniques in terrestrial HetVNets, unmanned aerial vehicle (UAV)-based techniques in air–ground vehicular networks, and satellite-based techniques in space–air–ground vehicular networks are investigated to off-load the cellular access networks and improve the content delivery performance. Second, existing works on mobile edge caching-assisted content delivery for backhaul offloading are further investigated, including the content placement scheme design and content delivery scheme design in different HetVNet scenarios.

2.1 HetVNet Techniques to Off-load Access Networks

To alleviate the burden caused by high vehicular communication demands, HetVNet-based data offloading (or traffic offloading) is an effective approach by utilizing complementary and revolutionary networking techniques to deliver mobile data originally planned for cellular transmissions. Basically, the HetVNet can be classified into two categories, i.e., a multi-tier network with single RAT and a multi-tier network with multiple RATs. In single-RAT HetVNet scenarios, the content delivery can be off-loaded to small cells like pico or femto cells, which also operate on the same cellular band as the macro BSs. The multi-RAT HetVNet, on the other hand, indicates that the cellular macrocells cooperate with other RATs, e.g., Wi-Fi and TVWS. Regarding the capacity constraint in cellular networks, we mainly target multi-RAT HetVNet for vehicular content delivery in this monograph.

2.1.1 Vehicular Traffic Offloading in the Terrestrial HetVNet

The HetVNet has emerged as one of the most promising network architectures to increase system throughput in wireless networks. Up to now, substantial heterogeneous offloading trials have been implemented in academic and industrial communities. In this part, we review the state-of-the-art data offloading strategies through heterogeneous terrestrial APs to enhance the vehicular content delivery performance. In particular, two candidate techniques, i.e., the Wi-Fi-based and TVWS-based techniques, are discussed in detail.

Wi-Fi-Based Techniques As a popular wireless broadband access technology, Wi-Fi operates on the unlicensed spectrum (2.4 and 5 GHz) and offers high data rates with limited coverage. In light of the cellular service performance degradation in overloaded areas, more and more network operators are extending their access networks by deploying Wi-Fi hotspots that are directly managed by them. For example, AT&T has deployed more than 30,000 Wi-Fi hotspots in the USA. China Mobile has deployed 4.4 million public Wi-Fi APs throughout China. Similarly, KT Corporation (formerly Korea Telecom) in South Korea owns and operates more than 140,000 Wi-Fi hotspots that are actively used for traffic offloading. Wi-Fi access is attractive because of the following advantages. (1) Wi-Fi hotspots are widely deployed in many urban areas. Globally, the total public Wi-Fi hotspots (including homespots) are forecasted to grow fourfold from 2018 to 2023, from 169 million in 2018 to 628 million by 2023 [1]. (2) Wi-Fi access is often free of charge or inexpensive. For example, China Mobile offers Wi-Fi services with less than $20 a month for unlimited data usage. (3) Most current mobile devices, such as smartphones, tablets, laptops, and more and more modern vehicles are equipped with Wi-Fi interfaces. (4) Currently, Wi-Fi technologies (e.g., IEEE 802.11ac/ad) can provide data rates of up to several Gbps.

In VNs, there exist many experimental and theoretical studies to evaluate the performance and demonstrate the effectiveness of Wi-Fi-based drive-thru networks. Based on the performance metrics, existing works can be classified into the following subcategories: *delay-oriented, throughput-oriented, and offloading efficiency-oriented.*

Delay-Oriented To evaluate the performance of Wi-Fi access delay (time required for authentication, IP address assignment, and so on), the authors in [2] adopt the Markov chain to study the impact of different factors on the access delay, such as the number of contending Wi-Fi users, wireless channel conditions, and the utilized authentication mechanisms. The accuracy of the theoretical analysis is then verified via MATLAB simulations and experimental testing. In [3], delay-constrained data transmission is investigated in Wi-Fi VNs, where data traffic is optimally distributed over cognitive radio and Wi-Fi interfaces to ensure timely and energy-efficient transmission.

Throughput-Oriented In addition to the delay, throughput is another important metric to evaluate the performance of Wi-Fi-assisted vehicular content delivery. In [4], the vehicular Wi-Fi throughput performance is investigated by considering the impact of the access procedure. Particularly, two access strategies, i.e., WPA2-PSK and Hotspot 2.0, are studied to show their impact on the throughput performance. In [5], focusing on the scenario where a Wi-Fi network coexists with an LTE network, the maximum total throughput is characterized considering two fairness constraints including throughput fairness and 3GPP fairness.

Offloading Efficiency-Oriented To improve the offloading efficiency, i.e., how much data can be off-loaded via Wi-Fi networks, a prediction-based offloading scheme named Wiffler is proposed in [6] to determine the accessed networks for different applications in HetVNets. Particularly, for delay-tolerant applications, Wiffler leverages a Wi-Fi connectivity prediction model to defer application data on Wi-Fi. For delay-sensitive applications, Wiffler proactively switches to cellular networks to avoid high delay penalties. In [7], a V2V-assisted Wi-Fi offloading scheme is proposed to improve the offloading efficiency in drive-thru Internet. Vehicles are associated with different APs and can assist peers' data tasks via V2V communications.

In addition to targeting only one performance metric in a work, there also exist some researches focusing on two or more metrics or exploring the trade-off among different metrics. For instance, in [8], the cost-effectiveness of a Wi-Fi solution for vehicular network access is investigated. By deploying Wi-Fi RSUs at the signalized intersection and studying the impact of traffic signals on Wi-Fi access, the trade-off between cost-effectiveness and the normalized service delay is studied. In [9], an adaptive joint Wi-Fi and cellular offloading scheme is proposed to flexibly adjust the trade-off between the energy consumption and the processing latency minimization, by considering factors including the task generation characteristics and channel access priorities for different users.

TVWS-Based Techniques Since analog TV broadcasting became obsolete, the TV spectrum has been significantly under-utilized currently. According to a study in Japan, more than $100\,\mathrm{MHz}$ of TVWS spectrum is observed available in about 84.3% of the country's area. Similarly, more than 50% of the TV channels are vacant in the USA and Hong Kong. More available TV spectrum is expected in some developing countries because of fewer TV stations used. Thanks to the analog-to-digital transition, which has been completed or is expected to be completed in the near future in many developed countries, a substantial amount of TV spectrum that was previously occupied by the TV broadcasting system can be released, which allows the unlicensed access when not utilized by licensed users. The unused TV spectrum is referred to as the TVWS band. Specifically, when operating in the TVWS band, to ensure non-interfering spectrum utilization, vehicles can only access TVWS channels that are not being used by incumbent users. In other words, the TVWS network access might be interrupted when primary users become active.

Compared with the widely adopted cellular and Wi-Fi radio interfaces, TVWS technologies have not been widely implemented. However, there have been many standards and research works focusing on unlicensed communications in the TVWS band. To better utilize the TVWS spectrum, regulators around the world have specified regulation frameworks for unlicensed access in the TVWS band [10]. Furthermore, a set of standards has also been proposed and adopted for unlicensed use of TVWS spectrum, including IEEE 802.11af, IEEE 802.22, and ETSI reconfigurable radio systems. IEEE 802.19.1 standard has also been published to facilitate the coexistence of heterogeneous networks in the TVWS band. Furthermore, there exist many industrial organizations (such as Carlson RuralConnect, Adaptrum, and 6Harmonics) providing devices and systems for TVWS Internet connectivity. Thus, it can be expected that the TVWS technology can also be widely implemented in the near future.

Substantial field tests and theoretical analysis have been done to evaluate the achievable performance of the TVWS communications. In [11], field tests as well as indoor evaluation of a multi-hop V2V communication system with distributed and autonomous TVWS channel selection are conducted in Japan. In [12], the performances of 802.11n (in 2.4 GHz) and 802.11af (in TVWS band) standards are compared in real environments by means of theoretical analysis and simulations. The results show that the 802.11n outperforms the 802.11af in terms of data rate but struggles in complex environments and NLOS conditions, where the 802.11af can effectively improve the communication performance. Focusing on TVWS as the offloading technology, a connection-aware balancing algorithm is proposed in [13] to exploit multiple radio connections to balance the load and route the traffic through the best possible interface given the network condition.

Recently there have been more and more research works focusing on exploiting TVWS for data offloading in VNs. In [14], the TVWS application scenarios in HetVNets are presented. The authors also propose two TVWS geolocation database-based architectures for V2V and V2I communications. The key technical challenges and future research directions toward exploiting TVWS for vehicular communications are also highlighted. In [15], the opportunistic usage of TVWS spectrum in VNs is investigated with different channel occupation policies. To improve the system throughput for V2V communications over TVWS band, a distributed joint channel and power allocation scheme is proposed in [16] by considering the spatio-temporal variations on TVWS channel availability and vehicular mobility.

To this end, it is widely recognized that the Wi-Fi and TVWS-based techniques can efficiently off-load the cellular traffic and improve the system performance. However, for data offloading through fixedly deployed APs, a very dense AP deployment is required to guarantee a uniform service coverage in the spatial domain. Furthermore, it will inevitably take a long period for hardware equipment upgrade with a high maintenance cost. Thus, more flexible and cost-effective traffic offloading techniques should be investigated to keep pace with the explosively increasing and unevenly distributed mobile data.

2.1.2 UAV-Assisted Traffic Offloading in the AGVN

The UAV-assisted traffic offloading is a promising solution to support vehicular content delivery in AGVNs. Compared with fixedly deployed APs, UAVs have several important advantages including flexible deployment, cost-effectiveness, and better LoS communication links. These benefits make UAV-aided wireless communications a promising integral component of future wireless systems to support diverse applications with orders-of-magnitude capacity improvement [17].

In recent years, UAVs have attracted increasing attention in industrial applications. For example, *General Atomics* and *Boeing* are among the notable manufactures providing military UAVs. *Prime Air* and *Project Wing* are delivery systems to offer rapid delivery by using UAVs. *AT&T* and *Verizon* have both conducted trials with LTE BSs mounted on UAVs. Moreover, *Qualcomm* is also planning to deploy UAVs for wireless communications in the 5G wireless networks. In academia, UAV-aided wireless communications have also gained increasing attention. In the following, we review the UAV-related research in two categories, i.e., UAV deployment and trajectory design and UAV-based AGVN design with known UAV mobility.

UAV Deployment and Trajectory Design in the AGVN The UAVs' mobility control is essential for exploiting the full potential of UAV-aided wireless communications. Therefore, in most current research works, the UAV-aided system is investigated with the optimization of UAV deployment or trajectory design to better serve the mobile users. The UAV deployment design aims to find the optimal locations for UAVs, including flying altitudes and horizontal positions, to achieve the best performance. In [18], a UAV-assisted HetVNet is investigated with a multi-layer vehicular architecture, and a density-aware deployment scheme is proposed to maximize the throughput with an iterative three-dimensional matching resource allocation algorithm. In [19], a three-dimensional UAV-cell deployment problem is investigated with a given number of UAVs being deployed to maximize the user coverage while maintaining UAV-to-user link qualities. Particularly, a per-UAV iterated particle swarm optimization algorithm is proposed to optimize UAV deployments for different UAV numbers.

Different from UAV deployment, the UAV trajectory design focuses on designing the optimal trajectory, following which the UAVs fly to serve users in multiple areas. Recently, UAV trajectory design has attracted increasing research attention due to its cost-effectiveness, since one flying UAV can cover multiple areas without requiring the deployment of multiple UAVs. Aiming to serve vehicles that are not covered by terrestrial infrastructure, a deep reinforcement learning (DRL)-based approach is proposed in [20] to optimize UAV trajectories by considering vehicular network dynamics. In [21], a multi-UAV trajectory planning and resource allocation (TPRA) problem is studied to maximize the accumulative network throughput while guaranteeing user fairness, UAV power consumption, and link quality constraints. In [22], a UAV is considered as a relay between BSs and vehicular users in the AGVN with optimized UAV trajectory and power allocation. UAV-aided computing offloading is investigated in [23], where a UAV is deployed to provide mobile edge computing services to a set of ground vehicles. An optimization framework for total utility maximization is developed by jointly optimizing the transmit power of vehicles and the UAV trajectory via a dynamic programming method.

UAV-Based AGVN Design with Known UAV Mobility In addition to the UAV deployment or trajectory optimization, there also exist many existing works on UAV-based AGVN design that leverage the full controllability of UAVs with the assumption that UAV positions/trajectories are known a priori. In [24], UAVs are utilized as store–carry–forward nodes to enhance the connection availability among vehicles and reduce the end-to-end packet delivery delay, where UAVs periodically broadcast their mobility information (including location, speed, and travel direction) to facilitate network management. In [25], an anti-jamming UAV relay problem is studied where UAVs relay messages from vehicles to RSUs to improve the communication performance against smart jammers. A hot-booting policy hill climbing-based UAV relay strategy is proposed to achieve the optimal relay policy without requiring information on the vehicular model and the jamming model. In [26], the real-time positions of vehicles and UAVs are assumed to be known by using a global positioning system (GPS). Then, an efficient routing scheme based on a flooding technique is proposed to ensure reliable and robust data delivery in the AGVN, where UAVs and vehicles cooperate in an ad hoc fashion to provide routing paths. UAVs in [27] act as relays to enhance the vehicular communication performance, where the mobility information of UAVs and vehicles is obtained via cooperative information exchange. The relay selection problem is then studied and formulated as a multi-objective optimization problem by jointly considering the state transition probability of communication interruption and the transmission energy and delay consumption.

2.1.3 Satellite-Assisted Traffic Offloading in the SAGVN

As the SAGVN is still in its infancy, some industrial and academic efforts have been devoted to construct satellite constellation systems and provide insights into the convergence of space, aerial, and terrestrial networks.

Industrial Efforts and Standardization Activities In the past decades, several LEO satellite constellation systems were established and applied for global wireless communications. The first global LEO satellite network is the *Iridium* system, mainly focusing on voice and low-rate data services. Other LEO satellite systems, including *Globalstar* and *Orbcomm*, tried to support satellite phone or Internet services to terrestrial users. However, these existing satellite networks did not perform well in the communication market and eventually went bankrupt due to the high construction cost. Recently, driven by the micro-satellite manufacturing and low-cost launch technologies, there are various initiatives to construct satellite constellations and launch thousands of LEO satellites. For instance, *Starlink* is an

LEO satellite system proposed by *SpaceX*, which plans to launch 42,000 satellites into orbits with altitudes of 340–1150 km. Currently, more than 1737 Starlink satellites have been launched, and the global service is expected by late 2021 or 2022 [28]. *OneWeb* satellite system is expected to include 648 LEO satellites on the orbits with an altitude of 1200 km, operating on the Ka and Ku bands. Since 2019, *OneWeb* has launched 110 satellites into the orbit, and the globally commercial usage is expected to begin in 2021 [29]. *TeleSat*, another LEO satellite network, is expected to have 300 satellites on the orbits with an altitude of 1000 km and provide global service starting in 2022 [30].

The standardization of satellite–terrestrial network integration has also been developed to guide the implementation of the SAGVN with high performance. The 3rd generation partnership project (3GPP) has done substantial standardization work on satellite-terrestrial network integration. Standards [31–33] have been developed to explore the scenarios, identify service requirements, and classify the service application scenarios of using satellite access in 5G. ETSI has also proposed some standards related to the convergence of satellite and terrestrial networks. In [34, 35], the definition and classification of communication scenarios, the role of satellites in disaster management, and the resource requirements for different applications are presented. In [36], the Satellite Independent Service Access Point (SI-SAP) is proposed and the physical air interfaces for broadband services are regulated.

In addition to industrial satellite constellation construction and the standardization processes, there are also some related projects exploring the solutions for the integrated SAGVN. Project *SANSA* (Shared Access Terrestrial-Satellite Backhaul Network enabled by Smart Antennas) proposes a satellite–terrestrial network to mitigate the backhaul pressure [37]. *VITAL* (VIrtualized hybrid satellite-TerrestriAl systems) project introduces SDN and network function virtualization (NFV) technologies to enable flexible management of the integrated network [38]. In project *SATNEX IV* (SATellite NEtwork of Experts IV), the utilization of terrestrial communication technologies in space networks is evaluated [39]. *SaT5G* (Satellite and terrestrial network for 5G) project aims to bring satellite communications into 5G by defining optimal satellite-based backhaul and traffic offloading solutions [40]. Project SATis5 aims to build a large-scale real-time live end-to-end 5G integrated satellite–terrestrial network proof-of-concept testbed [41].

Research Activities in the SAGVN The SAGVN shows great potential in the next-generation network systems. There have been a large number of works focusing on the possible architecture of the integrated SAGVN. In [42, 43], an SDN-based integrated network architecture is proposed to enable flexible, efficient, and global management for the SAGVN. In [44], the SAGVN is studied to address the issues including network reconfiguration under dynamic space resources constraints, multi-dimensional sensing and context information integration, and real-time and secure vehicular communications. In [45], a framework of SAG integrated moving cell, namely, SAGECELL, is proposed to combine space, aerial, and terrestrial networks in a complementary fashion for managing dynamic varying traffic demands with limited network resources.

Focusing on the communication resource allocation in the integrated network, a state-based and event-driven system model is proposed to facilitate content delivery in [46]. In [47], a cooperative multi-group multicast-based content delivery strategy is proposed. In particular, beamforming technologies are utilized to improve the network efficiency and to serve the diversified service requirements with limited radio resources. In [48], the ultra-dense LEO satellites are integrated with the terrestrial network to achieve efficient data offloading. In [49], an online control framework is proposed to make online decisions on the request admission and scheduling, UAV dispatching, and resource slicing for different services in the SAGVN.

To fully utilize the computing capability of heterogeneous devices in the SAGVN, edge computing enhanced SAGVN is promising to enable ubiquitous data processing and content sharing. In [50], the joint optimization of radio resource allocation and bidirectional communication and computation task offloading is investigated. The original optimization problem is then decoupled into two sub-problems and solved by the proposed heuristic algorithm. Focusing on the delay-oriented IoT services, a computing task scheduling problem is investigated in [51] to minimize the offloading and computing delay of all computing tasks. In [52], an SAG edge/cloud computing architecture is proposed to off-load the computation-intensive applications considering remote energy and computation constraints.

A comprehensive survey of the integrated SAG networks is provided in [53], where topics on cross-layer design, resource management and allocation, system integration, and network performance analysis are discussed. In [54], a comprehensive simulation platform is developed for the integrated SAG network, integrating multiple network protocols, node mobility, and control algorithms. Furthermore, various interfaces are provided to enable functionality extension to facilitate user-defined mobility traces and control algorithms. In view of the complex and dynamic network environment of the integrated SAG network, AI-based techniques are leveraged in [55] to improve the network performance by addressing challenges in network control, spectrum management, energy management, routing and handover management, and security guarantee.

Due to the merits of HetVNet techniques in supporting tremendous vehicular traffic demands, we will adopt the HetVNet architecture integrating terrestrial, aerial, and space networks in this monograph. Specifically, we will systematically investigate the impacts of diversified network characteristics on the content delivery performance. In addition, considering that current HetVNet-based mechanisms have rarely considered the high mobility of both vehicle users and APs (UAVs and LEO satellites), we will also investigate the mobility-aware content delivery scheme design to fully unleash the differential merits of heterogeneous network segments.

2.2   Caching-Assisted Content Delivery to Off-load Backhaul Networks

Mobile edge caching is a promising solution to address backhaul limitations including backhaul congestion, unreliable backhaul links, and long backhaul delay. Caching techniques enable storing content files in strategically placed caching nodes/servers during off-peak time, and redirecting content requests to the most appropriate servers at peak time [56–58]. In fact, caching-based content delivery networks (CDNs) have been increasingly deployed in the world. For example, *Akamai Intelligent Platform, Google Global Cache system, Facebook CDN, Amazon CloudFront*, and *ARA CDN* are several representative state-of-the-art caching systems. In this section, existing works on mobile edge caching are reviewed in two categories, i.e., content placement schemes and content delivery schemes.

2.2.1   Content Placement Schemes

The key idea of content placement is to cache content files in nodes close to the requesting users, enabling content download from caching nodes instead of the remote content server. To efficiently utilize the limited caching and communication resources, caching schemes should be designed specifically for different scenarios to support vehicular content delivery. Based on different caching decision-making manners, the caching schemes can be divided into different categories (such as proactive/reactive, centralized/distributed, deterministic/probabilistic) with their corresponding advantages and disadvantages, as summarized in Table 2.1. In the following, existing works on content placement in heterogeneous terrestrial APs, UAVs, and satellites are, respectively, reviewed.

Content Placement in Terrestrial HetVNets Focusing on distributed content storage in V2V scenarios, an efficient V2V-based caching strategy is investigated in [60], where each vehicle makes caching decisions independently by considering the diversified application requirements, the crucial features of data, and a set of key attributes of the VNs. In [61], an SDN-based incentive V2V caching scheme is proposed, where a small BS encourages vehicles to cache the popular content by offering them a reward. The single leader multiple followers Stackelberg game is utilized to model the problem and the Stackelberg equilibrium is derived. Focusing on V2I communications, content caching in roadside APs is investigated to maximize content retrievability while considering the limited storage capacities by proposing an ILP-based optimization framework in [62] and multi-object auction-based solutions in [63]. Allowing content caching in both vehicles and fixed APs, an edge caching scheme in RSUs is developed in [64] to analyze the vehicular content requests and determine where to obtain the requested content by considering the cooperation between vehicles and RSUs. In [65], assuming that content files can

Table 2.1 Content caching solutions summary [59]

Approach	Features	Advantages	Disadvantages
Reactive caching	Cache content after being requested	Only cache content that is actually requested; cost-effective	Additional overhead is required in the initial response
Proactive caching	Cache content before being requested	Alleviate peak-hour traffic and reduce network latency	Caching performance relies on prediction accuracy
Centralized caching	Decisions are made by a centralized entity	Achieve better network performance with a global network vision	Single point of failure; does not scale well
Distributed caching	Decisions are made distributedly with localized information	Does not require a global processing unit; easy access to up-to-date local information	May not get the global optimal solutions; high complexity of distributed algorithms and protocols
Deterministic caching	Content caching is determined based on given information	Better performance in static networks	Not suitable for dynamic situations
Probabilistic caching	Cache content with certain probabilities	Adaptive to uncertain network status	Hard to obtain the optimal solutions
Non-cooperative caching	Content caching is decided independently	Low complexity and signaling overhead	May result in inefficient cache utilization
Cooperative caching	Cache nodes share content with each other	High cache utilization efficiency	High signal overhead by sharing caching status; additional content retrieving delay
Uncoded caching	Complete or a part of a file is cached	Easy implementation; less processing required	Low cache utilization efficiency
Coded caching	Encode content into packets for caching	Improved reliability; high cache utilization efficiency	Require more computation resources

be cached at a macrocell BS, RSUs, and smart vehicles, the authors propose a cooperative edge caching scheme to jointly optimize the content placement and delivery. The optimization problem is then formulated as a double timescale Markov decision process and solved by the proposed deep deterministic policy gradient-based solutions.

Content Placement in UAVs Recently, content caching in UAVs has attracted wide research interest due to UAVs' inherent characteristics of limited wireless backhaul. Most of the current UAV-aided caching solutions focus on mobile networks with no or low user mobility rather than the highly dynamic VNs. However, the ideas behind these solutions are valuable to future related researches in VNs. Therefore, some caching strategies in non-vehicular networks are still analyzed below. In

[66], the framework of caching UAV-enabled small cell networks is proposed to alleviate backhaul congestion, reduce energy consumption, and improve the quality of experience (QoE). In [67], the content placement and the proactive deployment of caching-enabled UAVs are studied to optimize the users' QoE by considering the predicted content request distribution, user mobility pattern, and user–UAV associations. In [68], the positions of cache-enabled UAVs are optimized by first determining the best height to maximize the coverage area and then obtaining the optimal 2D position by enumeration search. Then, the caching probability optimization is formulated as a linear problem and solved with Lagrangian function. The joint UAV placement and caching problem is also studied in [69] to maximize the cache hit ratio by optimizing the deployment of UAVs, the content caching, and the UAV–user associations. Aiming to maximize the spectral efficiency, a hybrid caching strategy is proposed in [70], where a popularity threshold is optimized to divide content files into two subsets, i.e., popular files that are cached at all the UAVs and less popular files that are cached at only one UAV.

Content Placement in Satellites Content caching in satellites was initially proposed based on proxy services, where the content is cached in ground stations with the assistance of satellites. A representative is the cache satellite distribution system (CSDS) [71] in which the proxy caches periodically report to a central ground station about requests received from their clients. The central station then utilizes the satellite broadcast capability to push some Web documents to the participating proxy caches, which caches the documents locally for future local requests. Recently, content placement in satellite networks is primarily considered to be performed with on-board cache [72–75]. In [72], content placement in LEO satellites is optimized by considering the interactions among distributed satellites for individual content caching decision-making. An exchange-stable matching algorithm is proposed to solve the content caching problem based on a many-to-many matching game with externalities. Content caching in satellite–terrestrial networks is considered in [73], where a two-layer caching model is proposed to enable caching in both ground stations and satellites. To minimize the satellite bandwidth consumption, content requests during an aggregation window will be aggregated and then served by the satellite as the window expires. In [74], a multi-layered satellite network is considered, where content caching in LEO and GEO satellites is optimized to realize load balancing. In [75], a cognitive radio (CR)-based satellite–terrestrial network is considered and the content placement on different satellites is studied to optimize the cache hit rate based on content popularity. Furthermore, the successful download probability is analyzed and two optimal cache space assignment strategies are proposed based on different terrestrial user densities.

2.2.2   Content Delivery Schemes

The design of content delivery strategies aims to effectively and efficiently dissem-inate requested content files from caching nodes to requesting users. Generally, content delivery solutions can be classified into four different categories [76]: *reverse request path, content announcements, periodic broadcast, and delivery scheduling*. In *reverse request path* schemes, content requests are sent until reaching a caching node that has the required content, and then the requested content is delivered to the requesting node using the reverse path. In *content announcement* schemes, caching nodes announce to their neighbors the content files they have cached, and then vehicles interested in these content files can request directly from the caching nodes. *Periodic broadcast* schemes refer to solutions in which caching nodes periodically and actively broadcast content to the passing vehicles. The *delivery scheduling* schemes usually leverage the expected network topology to schedule a delivery path from caching nodes to requesting nodes by considering the expected contact between nodes. In the following, we review the content delivery strategies in HetVNet scenarios with different network types.

Content Delivery in Terrestrial HetVNets   In the terrestrial HetVNet, content files are delivered by V2V communications and/or V2I communications. Focusing only on V2V communications, a cached data transferring scheme is studied in [77] to maintain the stored data in a designated region. Particularly, to avoid the loss of cached data when caching vehicles leave the region of interests, leaving vehicles push their cached coded data packets to the incoming vehicles via the one-hop V2V link at the entrance/exit of the region. In [78], parked vehicles are utilized to increase the storage capacity of the content server by caching content files in parked vehicles. The authors propose a reverse request path-based solution, in which moving vehicles send their interests to other vehicles in a social spot (with parked vehicles), and then the content server sends the content if it is available in the cache. Considering both V2I and V2V communications, the authors propose an in-network caching scheme in VNs in [79], where every vehicle is not only a requester but also a cache node to serve other vehicles' requests. It is assumed that all the vehicles are interested in the same file. Thus the RSU first downloads the file from the server and then broadcasts the file to all the vehicles. Then vehicles receiving the file can cache this file and help spread it to other vehicles that have not received the file. In [80], the authors consider a scenario where vehicles can only download several packets of the entire file which is broadcast by the RSU. The authors propose a coalition formation algorithm to enable cooperative V2V content sharing among different vehicles to complement the missing packets.

Content Delivery in HetVNets with UAVs   When it comes to UAV-aided and satellite-aided content delivery algorithms, most related research works focus on mobile networks with low mobility rather than VNs. However, some of these studies are still discussed in the following because these solutions can provide insight into their applicability for VNs. In [81], the UAV-assisted caching is investigated in the AGVN, where UAVs proactively broadcast popular content to vehicles and the ground RSUs provide services on demand through unicast. In [82], content files are delivered by ground CBSs or UAVs upon requested. Particularly, the authors propose a cooperative UAV clustering scheme to form UAV clusters for cooperatively delivering the content to ground users and offloading traffic from cellular networks. A RaptorQ-based content dissemination mechanism is proposed in [83], in which UAVs are constantly broadcasting content files to ground moving vehicles. Particularly, content files are encoded by using RaptorQ codes to enhance caching performance and energy efficiency. In [84], a centralized UAV-based content delivery scheduling problem is investigated in which a control center is responsible for managing the service delivery. When the control center receives a service request, it will distribute the service request to different regions with different UAVs by considering the distance of delivery, region size, and user's priority. In [85], a cooperative framework is proposed where caching-enabled UAVs and RSUs cooperate to support vehicular content delivery via scheduling and content management.

Content Delivery in HetVNets with Satellites Focusing on context-aware multimedia content delivery over cooperative satellite–terrestrial networks, the authors of [86] point out the potential challenges caused by the inherently different network characteristics and propose a dynamic spectrum allocation scheme to ably provide context-aware content files. In [87], the multimedia content delivery over a GEO satellite–terrestrial network is investigated, where multicast multimedia content transmission is optimized via managing the radio resources. In specific, a multicast subgrouping-maximum satisfaction index algorithm is proposed, where the users are divided into multiple multicast subgroups based on the experienced channel qualities and radio resources are assigned based on subgroup configuration. In [88], content delivery in high-speed railways over a satellite-terrestrial network is studied. A scheduling and resource allocation algorithm is then proposed to enhance content delivery performance with the prediction of handovers and channel state information.

In this monograph, we will investigate the content caching and delivery scheme design in the SAGVN considering the impact of differentiated network characteristics, high mobility of vehicles/UAVs/satellites, content popularity, and APs' caching capacities. These inherent HetVNet characteristics have been rarely considered in existing works on HetVNet caching and delivery but essential for efficient vehicular service provisioning.

2.3 Summary

In this chapter, we have surveyed the existing works on the HetVNet-based vehicular traffic offloading and mobile edge content caching and delivery strategies in terrestrial HetVNets, AGVNs, and SAGVNs to improve the vehicular content delivery performance. With the literature review, we have achieved a clearer understanding of the limitations and deficiencies in the current studies on this area and stated the corresponding technical motivations of this monograph.

References

1. Cisco, Cisco annual internet report (2018–2023). Accessed July 2021. https://www.cisco.com/c/en/us/solutions/collateral/executive-perspectives/annual-internet-report/white-paper-c11-741490.pdf
2. W. Xu, H.A. Omar, W. Zhuang, X. Shen, Delay analysis of in-vehicle internet access via on-road WiFi access points. IEEE Access **5**, 2736–2746 (2017)
3. S.-S. Tzeng, Y.-J. Lin, Delay-constrained data transmission with minimal energy consumption in cognitive radio/WiFi vehicular networks. Wirel. Pers. Commun. **107**(4), 1777–1797 (2019)
4. W. Xu, W. Shi, F. Lyu, H. Zhou, N. Cheng, X. Shen, Throughput analysis of vehicular internet access via roadside WiFi hotspot. IEEE Trans. Veh. Technol. **68**(4), 3980–3991 (2019)
5. X. Sun, L. Dai, Towards fair and efficient spectrum sharing between LTE and WiFi in unlicensed bands: fairness-constrained throughput maximization. IEEE Trans. Wirel. Commun. **19**(4), 2713–2727 (2020)
6. A. Balasubramanian, R. Mahajan, A. Venkataramani, Augmenting mobile 3G using WiFi: measurement, system design, and implementation, in *Proc. ACM MobiSys, San Francisco, CA*, June 2010
7. W. Xu, H. Wu, J. Chen, W. Shi, H. Zhou, N. Cheng, X. Shen, ViFi: vehicle-to-vehicle assisted traffic offloading via roadside WiFi networks, in *Proc. IEEE GLOBECOM 2018, Abu Dhabi*, Dec 2018
8. N. Lu, N. Cheng, N. Zhang, X. Shen, J.W. Mark, F. Bai, Wi-Fi hotspot at signalized intersection: cost-effectiveness for vehicular internet access. IEEE Trans. Veh. Technol. **65**(5), 3506–3518 (2016)
9. W. Fan, J. Han, L. Yao, F. Wu, Y. Liu, Latency-energy optimization for joint WiFi and cellular offloading in mobile edge computing networks. Comput. Netw. **181**, 107570 (2020)
10. H. Harada, White space communication systems: an overview of regulation, standardization and trial. IEICE Trans. Commun. **E97.B**(2), 261–274 (2014)
11. O. Altintas, Y. Ihara, H. Kremo, H. Tanaka, M. Ohtake, T. Fujii, C. Yoshimura, K. Ando, K. Tsukamoto, M. Tsuru, Y. Oie, Field tests and indoor emulation of distributed autonomous multi-hop vehicle-to-vehicle communications over TV white space, in *Proc. ACM 18th Annu. Int. Conf. on Mobile Comput. and Netw., Istanbul*, Aug 2012, pp. 439–442
12. L. Bedogni, M.D. Felice, F. Malabocchia, L. Bononi, Cognitive modulation and coding scheme adaptation for 802.11n and 802.11af networks, in *Proc. IEEE GLOBECOM 2014 Workshop, Austin, TX*, Dec 2014
13. L. Bedogni, A. Trotta, M.D. Felice, Y. Gao, X. Zhang, Q. Zhang, F. Malabocchia, L. Bononi, Dynamic adaptive video streaming on heterogeneous TVWS and Wi-Fi networks. IEEE/ACM Trans. Netw. **25**(6), 3253–3266 (2017)
14. H. Zhou, N. Zhang, Y. Bi, Q. Yu, X. Shen, D. Shan, F. Bai, TV white space enabled connected vehicle networks: challenges and solutions. IEEE Netw. **31**(3), 6–13 (2017)
15. A. Arteaga, S. Cespedes, C. Azurdia-Meza, Vehicular communications over TV white spaces in the presence of secondary users. IEEE Access **7**, 53496–53508 (2019)
16. S. Midya, A. Roy, K. Majumder, S. Phadikar, QoS aware distributed dynamic channel allocation for V2V communication in TVWS spectrum. Comput. Netw. **171**, 107126 (2020)
17. H. Wu, F. Lyu, C. Zhou, J. Chen, L. Wang, X. Shen, Optimal UAV caching and trajectory in aerial-assisted vehicular networks: a learning-based approach. IEEE J. Sel. Areas Commun. **38**(12), 2783–2797 (2020)

18. B. Wang, R. Zhang, C. Chen, X. Cheng, Y. Jin, Density-aware deployment with multi-layer UAV-V2X communication networks. IET Commun. **14**(16), 2709–2715 (2020)

19. W. Shi, J. Li, W. Xu, H. Zhou, N. Zhang, S. Zhang, X. Shen, Multiple drone-cell deployment analyses and optimization in drone assisted radio access networks. IEEE Access **6**, 12518–12529 (2018)

20. M.S. Shokry, D. Ebrahimi, C. Assi, S. Sharafeddine, A. Ghrayeb, Leveraging UAVs for coverage in cell-free vehicular networks: a deep reinforcement learning approach. IEEE Trans. Mobile Comput. **20**(9), 2835–2847 (2021)

21. W. Shi, J. Li, H. Wu, C. Zhou, N. Cheng, X. Shen, Drone-cell trajectory planning and resource allocation for highly mobile networks: a hierarchical DRL approach. IEEE Internet Things J. **8**(12), 9800–9813 (2020)

22. O. Abbasi, H. Yanikomeroglu, A. Ebrahimi, N.M. Yamchi, Trajectory design and power allocation for drone-assisted NR-V2X network with dynamic NOMA/OMA. IEEE Trans. Wirel. Commun. **19**(11), 7153–7168 (2020)

23. L. Zhang, Z. Zhao, Q. Wu, H. Zhao, H. Xu, X. Wu, Energy-aware dynamic resource allocation in UAV assisted mobile edge computing over social internet of vehicles. IEEE Access **6**, 56700–56715 (2018)

24. W. Fawaz, R. Atallah, C. Assi, M. Khabbaz, Unmanned aerial vehicles as store-carry-forward nodes for vehicular networks. IEEE Access **5**, 23710–23718 (2017)

25. L. Xiao, X. Lu, D. Xu, Y. Tang, L. Wang, W. Zhuang, UAV relay in VANETs against smart jamming with reinforcement learning. IEEE Trans. Veh. Technol. **67**(5), 4087–4097 (2018)

26. O.S. Oubbati, N. Chaib, A. Lakas, P. Lorenz, A. Rachedi, UAV-assisted supporting services connectivity in urban VANETs. IEEE Trans. Veh. Technol. **68**(4), 3944–3951 (2019)

27. Y. He, D. Zhai, Y. Jiang, R. Zhang, Relay selection for UAV-assisted urban vehicular ad hoc networks. IEEE Wirel. Commun. Lett. **9**(9), 1379–1383 (2020)

28. Starlink, Accessed July 2021. https://en.wikipedia.org/wiki/Starlink

29. OneWeb Home Page, Accessed July 2021. https://www.oneweb.world/launches

30. Telesat, Accessed July 2021. https://www.telesat.com/leo-satellites/

31. 3GPP, Study on scenarios and requirements for next generation access technologies, version 15.0.0, release 15. Standard TR 38.913, July 2018

32. 3GPP, Service requirements for the 5G system; stage 1, version 17.2.0, release 17. Standard TS 22.261, 3GPP, Mar 2020

33. 3GPP, Study on using satellite access in 5G; stage 1, version 16.0.0, release 16. Standard TR 22.822, 3GPP, July 2018

34. ETSI, Satellite Earth Stations and Systems (SES), Combined satellite and terrestrial networks scenarios, document v1.1.1. ETSI TR 103 124, July 2013

35. ETSI, Satellite Earth Stations and Systems (SES), Overview of present satellite emergency communications resources, v1.2.2. ETSI TR 102 641, Aug 2013

36. ETSI, Satellite Earth Stations and Systems (SES), Broadband satellite multimedia (BSM); common air interface specification; satellite independent service access point (SISAP) interface: primitives. v1.2.1. ETSI TS 102 357, May 2015

37. SANSA Home Page, Accessed July 2021. https://sansa-h2020.eu/

38. VIrtualized hybrid satellite-TerrestriAl systems for resilient and flexible future networks. Accessed July 2021. https://cordis.europa.eu/project/id/644843

39. SATNEX IV Home Page, Accessed July 2021. https://artes.esa.int/projects/satnex-iv

40. Sat 5G Home Page, Accessed July 2021. http://sat5g-project.eu/

41. SATis5 Home Page, Accessed July 2021. https://satis5.eurescom.eu/

42. N. Zhang, S. Zhang, P. Yang, O. Alhussein, W. Zhuang, X. Shen, Software defined space-air-ground integrated vehicular networks: challenges and solutions. IEEE Commun. Mag. **55**(7), 101–109 (2017)

43. H. Wu, J. Chen, C. Zhou, W. Shi, N. Cheng, W. Xu, W. Zhuang, X. Shen, Resource management in space-air-ground integrated vehicular networks: SDN control and AI algorithm design. IEEE Wirel. Commun. **27**(6), 52–60 (2020)

44. Z. Niu, X.S. Shen, Q. Zhang, Y. Tang, Space-air-ground integrated vehicular network for connected and automated vehicles: challenges and solutions. Intell. Converged Netw. **1**(2), 142–169 (2020)

45. Z. Zhou, J. Feng, C. Zhang, Z. Chang, Y. Zhang, K.M.S. Huq, SAGECELL: software-defined space-air-ground integrated moving cells. IEEE Commun. Mag. **56**(8), 92–99 (2018)

46. G. Gür, S. Kafiloğlu, Layered content delivery over satellite integrated cognitive radio networks. IEEE Wirel. Commun. Lett. **6**(3), 390–393 (2017)

47. X. Zhu, C. Jiang, L. Yin, L. Kuang, N. Ge, J. Lu, Cooperative multi-group multicast transmission in integrated terrestrial–satellite networks. IEEE J. Sel. Areas Commun. **36**(5), 981–992 (2018)

48. B. Di, H. Zhang, L. Song, Y. Li, G. Y. Li, Ultra-dense LEO: integrating terrestrial-satellite networks into 5G and beyond for data offloading. IEEE Trans. Wirel. Commun. **18**(1), 47–62 (2018)

49. F. Lyu, P. Yang, H. Wu, C. Zhou, J. Ren, Y. Zhang, X. Shen, Service-oriented dynamic resource slicing and optimization for space-air-ground integrated vehicular networks. IEEE Trans. Intell. Transp. Syst. (2021). https://doi.org/10.1109/TITS.2021.3070542

50. G. Wang, S. Zhou, Z. Niu, Radio resource allocation for bidirectional offloading in space-air-ground integrated vehicular network. J. Commun. Inf. Netw. **4**(4), 24–31 (2019)

51. C. Zhou, W. Wu, H. He, P. Yang, F. Lyu, N. Cheng, X. Shen, Deep reinforcement learning for delay-oriented IoT task scheduling in space-air-ground integrated network. IEEE Trans. Wirel. Commun. **20**(2), 911–925 (2021).

52. N. Cheng, F. Lyu, W. Quan, C. Zhou, H. He, W. Shi, X. Shen, Space/aerial-assisted computing offloading for IoT applications: a learning-based approach. IEEE J. Sel. Areas Commun. **37**(5), 1117–1129 (2019)

53. J. Liu, Y. Shi, Z.M. Fadlullah, N. Kato, Space-air-ground integrated network: a survey. IEEE Commun. Surv. Tutorials **20**(4), 2714–2741 (2018)

54. N. Cheng, W. Quan, W. Shi, H. Wu, Q. Ye, H. Zhou, W. Zhuang, X. Shen, B. Bai, A comprehensive simulation platform for space-air-ground integrated network. IEEE Wirel. Commun. **27**(1), 178–185 (2020)

55. N. Kato, Z.M. Fadlullah, F. Tang, B. Mao, S. Tani, A. Okamura, J. Liu, Optimizing space-air-ground integrated networks by artificial intelligence. IEEE Wirel. Commun. **26**(4), 140–147 (2019)

56. H. Wu, J. Chen, W. Xu, N. Cheng, W. Shi, L. Wang, X. Shen, Delay-minimized edge caching in heterogeneous vehicular networks: a matching-based approach. IEEE Trans. Wirel. Commun. **19**(10), 6409–6424 (2020)

57. J. Chen, H. Wu, P. Yang, F. Lyu, X. Shen, Cooperative edge caching with location-based and popular contents for vehicular networks. IEEE Trans. Veh. Technol. **69**(9), 10291–10305 (2020)

58. F. Lyu, J. Ren, N. Cheng, P. Yang, M. Li, Y. Zhang, X. Shen, LEAD: large-scale edge cache deployment based on spatio-temporal WiFi traffic statistics. IEEE Trans. Mobile Comput. **20**(8), 2607–2623 (2021)

59. J. Yao, T. Han, N. Ansari, On mobile edge caching. IEEE Commun. Surv. Tutorials **21**(3), 2525–2553 (2019)

60. D.D. Van, Q. Ai, Q. Liu, D.-T. Huynh, Efficient caching strategy in content-centric networking for vehicular ad-hoc network applications. IET Intell. Transp. Syst. **12**(7), 703–711 (2018)

61. A. Alioua, S. Simoud, S. Bourema, M. Khelifi, S.-M. Senouci, A Stackelberg game approach for incentive V2V caching in software-defined 5G-enabled VANET, in *Proc. IEEE Symp. on Comput. and Commun. (ISCC), Rennes* (IEEE, Piscataway, 2020)

62. G. Mauri, M. Gerla, F. Bruno, M. Cesana, G. Verticale, Optimal content prefetching in NDN vehicle-to-infrastructure scenario. IEEE Trans. Veh. Technol. **66**(3), 2513–2525 (2017)

63. Z. Hu, Z. Zheng, T. Wang, L. Song, X. Li, Roadside unit caching: auction-based storage allocation for multiple content providers. IEEE Trans. Wirel. Commun. **16**(10), 6321–6334 (2017)

64. Z. Su, Y. Hui, Q. Xu, T. Yang, J. Liu, Y. Jia, An edge caching scheme to distribute content in vehicular networks. IEEE Trans. Veh. Technol. **67**(6), 5346–5356 (2018)

65. G. Qiao, S. Leng, S. Maharjan, Y. Zhang, N. Ansari, Deep reinforcement learning for cooperative content caching in vehicular edge computing and networks. IEEE Internet Things J. **7**(1), 247–257 (2019)

66. N. Zhao, F.R. Yu, L. Fan, Y. Chen, J. Tang, A. Nallanathan, V.C. Leung, Caching unmanned aerial vehicle-enabled small-cell networks: employing energy-efficient methods that store and retrieve popular content. IEEE Veh. Technol. Mag. **14**(1), 71–79 (2019)

67. M. Chen, M. Mozaffari, W. Saad, C. Yin, M. Debbah, C.S. Hong, Caching in the sky: proactive deployment of cache-enabled unmanned aerial vehicles for optimized quality-of-experience. IEEE J. Sel. Areas Commun. **35**(5), 1046–1061 (2017)

68. B. Jiang, J. Yang, H. Xu, H. Song, G. Zheng, Multimedia data throughput maximization in internet-of-things system based on optimization of cache-enabled UAV. IEEE Internet Things J. **6**(2), 3525–3532 (2019)

69. E. Lakiotakis, P. Sermpezis, X. Dimitropoulos, Joint optimization of UAV placement and caching under battery constraints in UAV-aided small-cell networks, in *Proc. the ACM SIGCOMM 2019 Workshop on Mobile AirGround Edge Computing, Syst., Netw, Appl., Beijing, Aug 2019*

70. F. Zhou, N. Wang, G. Luo, L. Fan, W. Chen, Edge caching in multi-UAV-enabled radio access networks: 3D modeling and spectral efficiency optimization. IEEE Trans. Signal Inf. Process. Netw. **6**, 329–341 (2020)

71. A. Armon, H. Levy, Cache satellite distribution systems: modeling analysis, and efficient operation. IEEE J. Sel. Areas Commun. **22**(2), 218–228 (2004)

72. S. Liu, X. Hu, Y. Wang, G. Cui, W. Wang, Distributed caching based on matching game in LEO satellite constellation networks. IEEE Commun. Lett. **22**(2), 300–303 (2017)

73. H. Wu, J. Li, H. Lu, P. Hong, A two-layer caching model for content delivery services in satellite-terrestrial networks, in *Proc. IEEE global commun. conference (GLOBECOM)* (IEEE, Washington, DC, 2016)

74. E. Wang, H. Li, S. Zhang, Load balancing based on cache resource allocation in satellite networks. IEEE Access **7**, 56864–56879 (2019)

75. E. Wang, X. Lin, S. Zhang, Content placement based on utility function for satellite networks. IEEE Access **7**, 163150–163159 (2019)

76. F.A. Silva, A. Boukerche, T.R.M.B. Silva, L.B. Ruiz, E. Cerqueira, A.A.F. Loureiro, Vehicular networks: a new challenge for content-delivery-based applications. ACM Comput. Surv. **49**(1), 1–29 (2016)

77. B. Hu, L. Fang, X. Cheng, L. Yang, Vehicle-to-vehicle distributed storage in vehicular networks, in *Proc. IEEE ICC 2018, Kansas City, MO*, May 2018

78. Z. Su, Y. Hui, S. Guo, D2D-based content delivery with parked vehicles in vehicular social networks. IEEE Wirel. Commun. Lett. **23**(4), 90–95 (2016)

79. H. Tian, Y. Otsuka, M. Mohri, Y. Shiraishi, M. Morii, Leveraging in-network caching in vehicular network for content distribution. Int. J. Distrib. Sens. Netw. **12**(6) (2016). https://doi.org/10.1155/2016/8972950

80. T. Wang, L. Song, Z. Han, B. Jiao, Dynamic popular content distribution in vehicular networks using coalition formation games. IEEE J. Sel. Areas Commun. **31**(9), 538–547 (2013)

81. S. Zhang, W. Quan, J. Li, W. Shi, P. Yang, X. Shen, Air-ground integrated vehicular network slicing with content pushing and caching. IEEE J. Sel. Areas Commun. **36**(9), 2114–2127 (2018)

82. H. Wu, X. Tao, N. Zhang, X. Shen, Cooperative UAV cluster-assisted terrestrial cellular networks for ubiquitous coverage. IEEE J. Sel. Areas Commun. **36**(9), 2045–2058 (2018)

83. S. Ortiz, C.T. Calafate, J.-C. Cano, P. Manzoni, C.K. Toh, A UAV-based content delivery architecture for rural areas and future smart cities. IEEE Internet Comput. **23**(1), 29–36 (2018)

84. H. Zhang, S. Wei, W. Yu, G. Chen, D. Shen, K. Pham, Scheduling methods for unmanned aerial vehicle based delivery systems, in *2014 IEEE/AIAA 33rd Digital Avionics Systems Conference (DASC), Colorado Springs, CO*, Oct 2014

85. A. Al-Hilo, M. Samir, C. Assi, S. Sharafeddine, D. Ebrahimi, Cooperative content delivery in UAV-RSU assisted vehicular networks, in *Proc. ACM MobiCom Workshop on Drone Assisted Wireless Communications for 5G and Beyond, London* (2020)

86. Y. Kawamoto, Z.M. Fadlullah, H. Nishiyama, N. Kato, M. Toyoshima, Prospects and challenges of context-aware multimedia content delivery in cooperative satellite and terrestrial networks. IEEE Commun. Mag. **52**(6), 55–61 (2014)

87. G. Araniti, I. Bisio, M. De Sanctis, A. Orsino, J. Cosmas, Multimedia content delivery for emerging 5G-satellite networks. IEEE Trans. Broadcast **62**(1), 10–23 (2016)

88. X. Wang, H. Liy, W. Yao, T. Lany, Q. Wu, Content delivery for high-speed railway via integrated terrestrial-satellite networks, in *Proc. IEEE WCNC, Seoul* (IEEE, Piscataway, 2020)

Chapter 3
Delay-Minimized Mobile Edge Caching in the Terrestrial HetVNet

Abstract The caching-assisted heterogeneous vehicular networks (HetVNets) are envisioned as a promising solution to support the ever-increasing vehicular applications. In this chapter, we investigate content caching in terrestrial HetVNets where Wi-Fi roadside units (RSUs), TV white space (TVWS) stations, and cellular base stations (CBSs) are considered to cache content files. To characterize the intermittent Wi-Fi and TVWS network connections, we establish an on–off model with service interruptions to describe the vehicular content delivery process. Content coding is then leveraged to resist the impact of unstable network connections with optimized coding parameters. By jointly considering the impact of file profiles and network characteristics, we investigate the content placement in heterogeneous APs to minimize the average content delivery delay, which is formulated as an integer linear programming problem. Adopting the idea of the student admission model, the formulated problem is then transformed into a many-to-one matching problem and solved by our proposed stable matching-based caching scheme. Simulation results demonstrate that the proposed scheme can achieve near-optimal performances in terms of delivery delay and offloading ratio with low complexity.

The multifarious CAV applications with stringent QoS requirements have imposed significant challenges on current cellular networks. To alleviate the cellular network congestion problem and reduce the communication cost, it is imperative to explore alternative network resources to support vehicular content delivery [1]. Generally, different radio access technologies have their pros and cons, including transmission delay, throughput, jitter, etc. For example, Wi-Fi networks can significantly off-load cellular traffic in vehicular scenarios with properly designed offloading strategies [2], but the Wi-Fi connections are intermittent for vehicles due to the limited coverage and high vehicle mobility. Another potential solution to alleviate the cellular spectrum scarcity problem is to leverage the TVWS band. Due to its low frequency, the TVWS band can provide high penetration capabilities, low path loss, and wide coverage up to several kilometers [3, 4]. However, the TVWS network access may be interrupted by the primary users to ensure non-interfering spectrum access. Therefore, in this chapter, we focus on the terrestrial HetVNet where Wi-Fi and TVWS networks coexist with cellular networks to exploit their complementary advantages. In addition, mobile edge caching is a promising solution to facilitate content delivery by caching popular content in the HetVNet APs to relieve the backhaul traffic with a reduced delivery delay [5, 6]. Therefore, in this chapter, we focus on edge caching in terrestrial HetVNets to serve the vehicular content requests.

Although caching-assisted content delivery has attracted substantial research interests as mentioned in Chap. 2, the following problems, which are essential in the VNs to provide quality vehicular services, are insufficiently studied in existing works: (1) in the highly dynamic and unreliable VNs, content delivery might encounter service interruptions, which significantly affect the caching performance and further the caching policy design; however, this inherent vehicular characteristic has not been considered in existing works on HetVNet caching, (2) most existing works do not take full advantage of the heterogeneous network resources to boost the caching performance gain, and (3) the street layout or vehicle mobility patterns in most existing works are idealized or assumed to follow certain distributions, which is not practical.

In this chapter, we investigate content caching in terrestrial HetVNet APs, i.e., Wi-Fi RSUs, TVWS stations, and CBSs, to minimize the vehicular content delivery delay by taking into account the above-mentioned problems. Considering the high vehicle mobility and the intermittent network access provided by Wi-Fi and TVWS transmissions, the amount of data that can be transmitted within one coverage area is limited. Therefore, caching the whole content files, especially large files, in the Wi-Fi RSUs or TVWS stations is inefficient since the downloading of one file requires multiple encounters with Wi-Fi RSUs or TVWS stations. To resist the impact of intermittent network connections and enhance caching efficiency, coded caching is applied to encode files into small packets. Each AP only caches some encoded packets with small sizes. Specifically, fountain codes [7] are adopted in this chapter to encode files due to the high flexibility and reliability and low decoding complexity [8]. By modeling the intermittent Wi-Fi/TVWS network connections as on–off service processes, the delivery delay is analyzed by applying partial repeat-after-interruption (PRAI) transmission mode. Based on the delay analysis, the caching placement problem is formulated as an ILP problem. To solve the formulated problem efficiently with low complexity, the ILP problem is further transformed into a matching problem [9] between content files and HetVNet APs. Specifically, by designing the preference lists of content files and HetVNet APs based on file popularity, vehicle mobility, and APs' storage capacities, a Gale–Shapley (GS) [10] matching-based caching scheme is proposed to obtain a stable matching. The main contributions are listed as follows:

- By leveraging the interplay between file profiles and network characteristics, the problem of edge caching in heterogeneous APs is investigated. Specifically, the dynamics of the content files (e.g., file size and file popularity) and the network conditions (e.g., network capacity, AP distribution, and vehicle mobility pattern) are jointly considered, which facilitates efficient content caching schemes to achieve minimize the average delivery delay.

- We model the intermittent Wi-Fi and TVWS network connections as on–off service processes with generally distributed on-periods and off-periods. Furthermore, coded caching is leveraged to resist the impact of unstable network connections and the coding parameters are optimized to adapt to the characteristics of different access networks. Then, by applying PRAI transmission mode, the proposed coded caching scheme can achieve a good balance between delivery delay and offloading performance.
- The problem of content caching in HetVNets with service interruptions is formulated as a many-to-one matching problem and solved by our proposed stable matching-based caching scheme. The construction of the two-sided preference lists is multi-objective, considering both the delivery delay and offloading performances. With the carefully designed preferences for content files and the APs, our matching-based caching scheme achieves a good performance with a low time complexity.
- Extensive experiments are conducted to demonstrate the effectiveness of the proposed caching scheme, by comparing multiple performance metrics including delivery delay, offloading ratio, and cache hit rate.

The remainder of this chapter is organized as follows. In Sect. 3.1, the system model and problem formulation are provided. In Sect. 3.2, the detailed derivation of the average content delivery delay from HetVNet APs is analyzed, and the matching-based content placement optimization scheme is described in Sect. 3.3. Simulation results are presented in Sect. 3.4 and a brief summary of this chapter is provided in Sect. 3.5. Main notations used in this chapter are listed in Table 3.1.

3.1 System Scenario and Problem Formulation

3.1.1 Scenario Description

In this chapter, vehicular users are assumed to be equipped with three radio interfaces for cellular, Wi-Fi, and TVWS communications.[1] The communication scenario is shown in Fig. 3.1a. Focusing on urban/sub-urban scenarios where CBSs are densely deployed, we assume that CBSs can provide seamless network connections for vehicles. Content delivery from Wi-Fi RSUs or TVWS stations is available only when vehicles travel through the corresponding coverage areas, as shown in Fig. 3.1b. The intermittent Wi-Fi/TVWS network connections are modeled

---

[1] In addition to Wi-Fi- and TVWS-based access technologies, there exist many other techniques [11]. Although only Wi-Fi, TVWS, and cellular networks are considered in this chapter, our methodology and the proposed scheme are applicable to HetVNets scenarios with more access techniques.

Table 3.1 Summary of notations

r_T, r_W	Coverage radius of Wi-Fi RSUs and TVWS stations, respectively
$\overline{R}_T, \overline{R}_W, \overline{R}_C$	Average TVWS, Wi-Fi, and CBS transmission rates, respectively
k_m, α_m	The number of source (or encoded) packets and the size of each packet for file f_m, respectively
k_m	Number of source packets for file f_m
K_m	Number of encoded packets required to recover file f_m
p_{req}^m	Probability that file f_m is requested by vehicles
p_{suc}^W	Probability that vehicles can successfully download at least one encoded packet from Wi-Fi RSUs
$p_{\max}^W$	Probability that vehicles can download enough encoded packets from RSUs without wasting time
$p_{\text{on}}^W(x), p_{\text{off}}^W(x)$	Pmfs of the time length of the on-periods and off-periods for Wi-Fi content downloading
a_m^W, a_m^T, a_m^C	Indicators showing the caching of file f_m in Wi-Fi RSUs, TVWS stations, and CBSs, respectively
n_m^W, n_m^T	Number of encoded packets of f_m cached in each Wi-Fi RSU and TVWS station, respectively
$\overline{D}_m^W, \overline{D}_m^T, \overline{D}_m^C, \overline{D}_m^B$	Average download delay of f_m from Wi-Fi, TVWS, CBS, and backhaul delivery, respectively
C_T, C_W, C_C	Storage capacities of the Wi-Fi RSUs, TVWS stations, and CBSs, respectively
$\mu_{\text{on}}^W, \mu_{\text{off}}^W$	Average time length in slots for on- and off-periods for Wi-Fi transmission
σ	Probability that an arbitrary slot is an on-slot
δ	Probability that on-period continues after an on-slot
$\mathcal{P}_{\text{files}}(f_m, I), \mathcal{P}_I(I, f_m)$	File f_m's preference over the APs and APs' preference over content files

as on–off processes, which will be introduced in detail in Sect. 3.1.2. When a vehicle generates a content request, there are two possible cases as shown in Fig. 3.1c: (1) *cache miss*—if the requested file is not cached in the APs, it can be fetched from the CBSs via backhaul links, and (2) *cache hit*—the requested file is cached and the vehicle can get it directly from the APs based on the caching location.

The TVWS and Wi-Fi coverage radii are denoted by r_T and r_W, respectively. The average transmission data rate of a Wi-Fi RSU–vehicle link is denoted by $\overline{R}_W = R_W^a / N_W^a$, where R_W^a is the achievable aggregate rate and N_W^a is the number of vehicles associated with one Wi-Fi RSU. Likewise, the average TVWS and CBS transmission rates are denoted by $\overline{R}_T$ and $\overline{R}_C$, respectively. Let $\mathcal{F} = \{f_1, \ldots, f_M\}$ be the set of M files and $\mathbf{z}_f = [z_1, \ldots, z_M]$ be the vector representing the sizes of M content files that are potentially requested by vehicular users. Files in set $\mathcal{F}$ are sorted by descending order based on the file popularity.[2] Considering that the popularity

---

[2] File popularity information can be obtained based on historical requests and predicted as studied in many existing works (e.g., [12]). Popularity prediction is beyond the scope of this chapter.

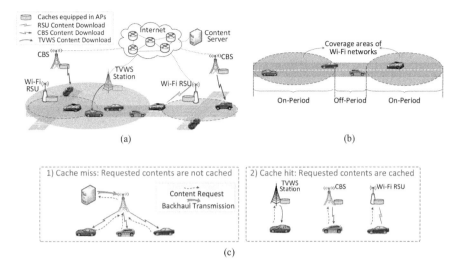

Fig. 3.1 Caching-based content delivery scenario in HetVNet. (**a**) Communication scenario with coexistence of cellular, Wi-Fi, and TVWS access networks. (**b**) An example of on–off service processes of Wi-Fi networks. (**c**) Content delivery cases: cache miss and cache hit

distribution of the network content items (e.g., YouTube videos) approximately follows Zipf's law, we model the file popularity by the Zipf distribution [13]. Thus, the request probability of f_m, which is the m-th popular file, is expressed as

$$p_{\text{req}}^m = \frac{1}{m^\xi} \bigg/ \left(\sum_{k=1}^M \frac{1}{k^\xi} \right), \tag{3.1}$$

where ξ ($\xi \geq 0$) controls the relative popularity of files, i.e., a larger ξ means that the first few popular files account for the majority of requests.

3.1.2 On–Off Service Model

Recall that Wi-Fi RSUs and TVWS stations provide intermittent network connections, which are modeled as on–off processes in this chapter. For Wi-Fi transmissions, as shown in Fig. 3.1b, *on-periods* correspond to the time duration when Wi-Fi access is available and *off-periods* appear when the vehicle is not covered by Wi-Fi RSUs. For TVWS transmissions, the off-periods also include the duration when TVWS channels are occupied by incumbent users and not available for secondary access. We consider a discrete-time system to divide on-periods and off-periods into constant length intervals called slots. Taking Wi-Fi transmission as an example, we define the length of one slot as the time required to transmit one bit of data by Wi-Fi RSUs: $l = 1/\overline{R}_W$. Thus, an on-period with time length of T_{on}^W has T_{on}^W/l slots.

Let $p_{on}^W(x)$ and $p_{off}^W(x)$ be the probability mass functions (pmfs) of the duration of on-periods and off-periods for the Wi-Fi transmission. Basically, the on–off period distributions are affected by the vehicle mobility patterns and the characteristics of RSUs including the deployment density and the coverage radius. The distributions of the on- and off-periods in a certain area can be obtained by observing the vehicle mobility traces. Alternatively, the distributions can also be assumed to follow geometrical distributions as [14] for analysis simplicity. In this chapter, the on- and off-periods are assumed to be generally distributed, and the proposed scheme can be applied to general cases with any known distributions for the on- and off-periods.

3.1.3 Fountain Coding

In this chapter, fountain codes are used to encode files cached in Wi-Fi RSUs and TVWS stations. Specifically, LT (Luby Transform) codes [15], the first proposed fountain codes, are adopted and will be briefly introduced in the following.

With LT codes, a source file f_m is partitioned into k_m source packets $s_1, s_2, \ldots, s_{k_m}$, each of which has a size of $\alpha_m = \frac{z_m}{k_m}$, where z_m is the total size of f_m. Then, encoded packets are obtained from the bitwise exclusive-or (XOR) of d randomly selected source packets, where d follows a degree probability distribution $\Omega(d)$ with $1 \leq d \leq k_m$. In other words, with d obtained from $\Omega(d)$, a vector $(v_1, v_2, \ldots, v_{k_m})$ is generated randomly satisfying $v_i \in \{0, 1\}$ for $i = 1, 2, \ldots, k_m$ and $\sum_{i=1}^{i=k_m} v_i = d$. The encoded packet is $\sum_{i=1}^{i=k_m} v_i s_i$ (bitwise sum modulo 2). From any set of K_m encoded packets (K_m is slightly larger than k_m, which will be explained at the end of this subsection), source file f_m can be recovered and decoded with success probability $1 - \epsilon$, where ϵ is the decoding failure probability when receiving K_m encoded packets.

Following [15], the degree distribution $\Omega(d)$ follows the *Robust Soliton Distribution*. Let $R \equiv c\sqrt{k_m}\ln(\frac{k_m}{\epsilon})$ with constant parameters $c > 0$ and $0 < \epsilon \leq 1$. Define

$$\rho(d) = \begin{cases} 1/k_m & \text{for } d = 1 \\ 1/[d(d-1)] & \text{for } d = 2, \ldots, k_m \end{cases}, \quad (3.2)$$

$$\phi(d) = \begin{cases} R/(k_m d) & \text{for } d = 1, \ldots, \frac{k_m}{R} - 1 \\ R\ln(R/\epsilon)/k_m & \text{for } d = \frac{k_m}{R} \\ 0 & \text{for } d > \frac{k_m}{R} \end{cases},$$

$$\beta_m = \sum_{d=1}^{d=k_m} [\rho(d) + \phi(d)].$$

Then we have $\Omega(d) = [\phi(d) + \rho(d)]/\beta_m$ for $d = 1, \ldots, k_m$. To guarantee a successful decoding probability of no smaller than $1 - \epsilon$, at least $K_m = k_m \beta_m$ encoded packets are required. Since $\sum_d \rho(d) = 1$, β_m is always larger than 1. Thus, the improvement of caching reliability in Wi-Fi RSUs and TVWS stations is achieved at the cost of total storage space consumption. Considering that CBSs are always available for content delivery, files cached in CBSs are stored without coding to avoid unnecessary extra storage occupancy and delivery delay.

3.1.4 Problem Formulation

Our objective is to optimize the content caching in HetVNet APs to minimize the average content delivery delay. Basically, caching files with various popularities and data sizes in different types of APs leads to different content delivery performances. Therefore, we jointly consider the file profiles and network characteristics to facilitate efficient content caching scheme design in heterogeneous terrestrial APs to minimize the average delivery delay. Let a_m^W, a_m^T, and a_m^C denote the caching indicator of file f_m in Wi-Fi RSUs, TVWS stations, and CBSs, respectively, where

$$a_m^W = \begin{cases} 1, & \text{file } f_m \text{ is cached in Wi-Fi RSUs} \\ 0, & \text{Otherwise,} \end{cases}$$

$$a_m^T = \begin{cases} 1, & \text{file } f_m \text{ is cached in TVWS stations} \\ 0, & \text{Otherwise,} \end{cases}$$

$$a_m^C = \begin{cases} 1, & \text{file } f_m \text{ is cached in CBSs} \\ 0, & \text{Otherwise.} \end{cases}$$

Notice that one content file can be cached in only one type of APs for the following reasons. (1) When adopting encoded caching, if a vehicle downloads encoded packets of a file from multiple access networks, the HetVNet APs need to negotiate and keep the same coding parameters to ensure successful content decoding. However, different access networks are managed by different service operators, which are competitors in the market and generally do not cooperate. (2) By constraining $a_m^W + a_m^T + a_m^C \leq 1$, the storage efficiency can be improved and more files can be cached, which facilitates the overall content delivery delay minimization.

Considering the limited AP storage capacities, the content caching should be carefully optimized. Specifically, the problem of content caching in HetVNet APs to minimize the overall average delivery latency can be formulated as

$$\min_{\mathbf{A}_T, \mathbf{A}_W, \mathbf{A}_C} \sum_{m=1}^{M} p_{\text{req}}^{m} \left(a_m^W \overline{D}_m^W + a_m^T \overline{D}_m^T + a_m^C \overline{D}_m^C + \left(1 - a_m^T - a_m^W - a_m^C \right) \overline{D}_m^B \right)$$

$$\tag{3.3}$$

$$s.t. \quad \sum_{m=1}^{M} a_m^T n_m^T \alpha_m \le C_T, \tag{3.3a}$$

$$\sum_{m=1}^{M} a_m^W n_m^W \alpha_m \le C_W, \tag{3.3b}$$

$$\sum_{m=1}^{M} a_m^C z_m \le C_C, \tag{3.3c}$$

$$a_m^W + a_m^T + a_m^C \le 1, \forall m = 1, \dots, M, \tag{3.3d}$$

$$a_m^W, a_m^T, a_m^C \in \{0, 1\}, \tag{3.3e}$$

where $\mathbf{A}_W = \left[a_1^W, \dots, a_m^W, \dots, a_M^W \right]$, $\mathbf{A}_T = \left[a_1^T, \dots, a_m^T, \dots, a_M^T \right]$, and $\mathbf{A}_C = \left[a_1^C, \dots, a_m^C, \dots, a_M^C \right]$. C_W, C_T, and C_C are the storage capacities of Wi-Fi RSUs, TVWS stations, and CBSs, respectively. n_m^T and n_m^W denote the numbers of encoded packets of file f_m cached in each TVWS station and Wi-Fi RSU, respectively. $\overline{D}_m^W$, $\overline{D}_m^T$, $\overline{D}_m^C$, and $\overline{D}_m^B$ denote the average delays of downloading file f_m from the Wi-Fi, TVWS, CBS, and backhaul transmissions, respectively. Constraints (3.3a)–(3.3c) indicate that the size of files cached in the APs should not exceed the corresponding maximum storage capacities. Constraint (3.3d) means that one content file can be cached in at most one type of APs.

3.2 Average Delivery Delay Analysis in HetVNet

To design a caching policy with minimized average content delivery delay, the delay performances of different delivery options (i.e., Wi-Fi, TVWS, CBS, and backhaul transmissions) should be analyzed. Firstly, for files encoded and cached in Wi-Fi RSUs and TVWS stations, coding parameters are optimized based on the file profiles and network characteristics. The PRAI transmission mode is then adopted to deliver encoded packets, and the corresponding average delivery delay is analyzed. In addition, the delays of the CBS and backhaul transmissions will also be analyzed.

3.2.1 Determination of Coding Parameters

Affected by the AP deployment and vehicle mobility patterns, the distributions of the on- and off-periods for Wi-Fi and TVWS transmissions are spatially and temporally variant. For instance, the on-periods in urban scenarios generally sustain longer than in rural areas due to lower vehicle velocity and denser AP deployment.

Focusing on urban scenarios, the on–off period distributions vary in rush hours and in off-peak hours due to different vehicle densities and velocities. The information on these distributions, however, can be obtained and regularity can be observed by monitoring vehicle mobility traces over a certain area. Without loss of generality, in this chapter, we assume that the characteristics of the on–off service process are known in advance. In the following, content download from Wi-Fi RSUs is taken as an example to illustrate the impact of the on–off model on the determination of coding parameters.

Basically, the time length of the on-periods determines the amount of data that can be transmitted within one coverage area. In our coding-based caching scheme, the size of one encoded packet is determined based on the distribution of on-periods. To ensure that vehicles traveling through a Wi-Fi coverage area can successfully download at least one encoded packet with a probability of at least p_{suc}^{W}, we determine the coding parameters k_m^{W} and α_m^{W} based on

$$\Pr(T_{on}^{W} \geq \alpha_m^{W}/\overline{R}_W) \geq p_{suc}^{W} \quad \Rightarrow \quad \sum_{T_{on}^{W}=\alpha_m^{W}/\overline{R}_W}^{\infty} p_{on}^{W}(T_{on}^{W}) \geq p_{suc}^{W}, \qquad (3.4)$$

where T_{on}^{W} is the length of the on-period.

With any known distribution of the on-periods, we can obtain the upper bound for the value of α_m^{W} from (3.4), which is denoted by α_m^{max}. Considering that the encoding and decoding complexities of LT codes increase with k_m^{W} [8], we choose the smallest possible value of k_m^{W} (the largest possible value of α_m^{W}) as follows:

$$k_m^{W} = \lceil z_m/\alpha_m^{max} \rceil, \quad \alpha_m^{W} = z_m/k_m^{W}. \qquad (3.5)$$

Considering the high vehicle mobility, the amount of data that can be downloaded by vehicles within each RSU varies from one another. Therefore, the number of packets cached in each RSU should be carefully designed. On the one hand, a small number of cached encoded packets may render large delivery delay for vehicles spending a long time in the coverage area, as they have to wait after downloading all the cached packets in the RSU. On the other hand, it may lead to a low caching storage efficiency if too many packets are cached in each RSU, while vehicles can never download so much data within one coverage area.

To achieve a good trade-off between delivery delay and storage efficiency, the number of encoded packets cached in each RSU can be determined based on service requirements. For instance, to ensure that at least $p_{max}^{W} \times 100\%$ of the vehicles can download enough number of packets for f_m within one RSU without wasting time, each RSU should cache at least n_m^{W} packets:

$$\Pr\left(0 \leq T_{on}^{W} \leq \frac{n_m^{W}\alpha_m^{W}}{\overline{R}_W}\right) \geq p_{max}^{W} \quad \Rightarrow \quad \sum_{0}^{\frac{n_m^{W}\alpha_m^{W}}{\overline{R}_W}} p_{on}^{W}(T_{on}^{W}) \geq p_{max}^{W}. \qquad (3.6)$$

Algorithm 1: Determination of coding parameters for Wi-Fi transmission

---

$\mathcal{F}$: Set of all content files. K_m^W: Number of encoded packets required to decode file f_m.
z_m: Size of file f_m. n_m^W: Number of encoded packets cached in each Wi-Fi RSU.
α_m: Size of one encoded packet. k_m^W: Number of source packets for file f_m.
begin

    for $f_m \in \mathcal{F}$ **do**

        Calculate the coding parameters α_m^W and k_m^W based on Eqs. (3.4)–(3.5).

        if $k_m^W = 1$ *and* $\alpha_m^W = z_m$ **then**

            File f_m has a small file size and can be cached in Wi-Fi RSUs without coding.
            $n_m^W = 1, K_m^W = 1$.

        else

            Obtain the value of n_m^W based on Eq. (3.6) and calculate the value of K_m^W
            based on analysis in Sect. 3.1.3. Set $n_m^W = \min\{n_m^W, K_m^W\}$.

        end

    end

    Output: α_m^W, k_m^W, n_m^W, and K_m^W for any $f_m \in \mathcal{F}$.

end

With known pmf for on-periods, we can easily obtain the values of α_m^W, k_m^W, K_m^W, and n_m^W. Note that small files (with $k_m^W = 1$ and $\alpha_m^W = z_m$) can be cached without coding to avoid extra storage occupancy and delivery delay. In addition, when the value of n_m^W calculated from Eq. (3.6) is larger than K_m, then $n_m^W = K_m^W$ since K_m^W encoded packets are sufficient to recover f_m. The detailed procedure of coding parameter determination is given in Algorithm 1. Similar analysis can be applied to coded caching parameter design in TVWS stations.

3.2.2 Effective Service Time

First, we define the terms *service time* and *effective service time (EST)*. Service time of a packet is the time required for transmission without interruption. Defining the length of one slot as the time required to transmit one bit by Wi-Fi RSUs, the service time of f_m in Wi-Fi transmission is equal to z_m slots. On the other hand, the EST to deliver file f_m is the time duration between the slot of request generation and the end of the slot when K_m encoded packets have been transmitted. Thus, the EST includes the periods when content downloading is available and the time duration when the service is interrupted. In the following, the EST of content delivery is analyzed by taking the Wi-Fi content delivery as an example.

Recall that LT-based encoded caching is utilized for Wi-Fi content caching. Thus, continuous-after-interruption (CAI) transmission mode is not suitable since the encoded packets cached in different Wi-Fi RSUs are different. Thus, PRAI transmission mode is adopted in this chapter. Specifically, if the transmission of an encoded packet is interrupted, the packet will be dropped and a new encoded packet of the same content file needs to be transmitted when the vehicle encounters

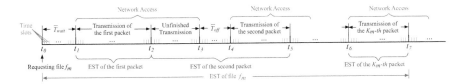

Fig. 3.2 Effective service time illustration for PRAI transmission mode

another Wi-Fi network access. Taking the illustration in Fig. 3.2 as an example, a vehicle requests file f_m with size z_m at time t_0 and content delivery starts as soon as the network access is available at time t_1. In other words, if the vehicle is not covered by a Wi-Fi RSU when generating the file request, it has to wait for $t_1 - t_0$ to get served; otherwise, $t_1 = t_0$. After successfully downloading the first packet of f_m, the transmission of the second packet is interrupted when the vehicle moves out of the Wi-Fi coverage area. When entering the coverage area of another Wi-Fi RSU, the second packet (not necessarily the same as the unfinished one) needs to be retransmitted. Thus, the EST of the second packet is $t_5 - t_2$. Correspondingly, the EST of file f_m is $t_7 - t_0$, which includes the waiting period $t_0 \sim t_1$ and the ESTs of K_m packets.

3.2.3 Average Delivery Delay Analysis

Let μ_{on}^W and μ_{off}^W denote the average number of slots for on- and off-periods. Denote by σ the probability that an arbitrary slot is an on-slot, which can be expressed as

$$\sigma = \mu_{on}^W / \left(\mu_{on}^W + \mu_{off}^W \right). \tag{3.7}$$

If a vehicle is not covered by the Wi-Fi network when generating the file request, it has to wait for a certain time to get served. Let T_{off}^W denote the duration of the off-period in slots with mean μ_{off}^W. Thus, the average number of slots for the waiting time is

$$\overline{T}_{wait}^W = E \left\{ (1 - \sigma) \cdot \sum_{x=1}^{T_{off}^W} \frac{1}{T_{off}^W} x \right\} \tag{3.8}$$

$$= (1 - \sigma) \cdot E \left\{ \frac{1}{T_{off}^W} \cdot \frac{T_{off}^W (T_{off}^W + 1)}{2} \right\} = \frac{(1 - \sigma)}{2} (\mu_{off}^W + 1).$$

Denote by $\mathbb{A}$ the event that the on-period continues after an on-slot. Let δ denote the probability that event $\mathbb{A}$ happens and T_{on}^W denote the duration of on-periods with mean value μ_{on}^W. We have

$$\delta = \sum_{x=0}^{\infty} \Pr(\mathbb{A}|T_{\text{on}}^{W} = x) \times \Pr(T_{\text{on}}^{W} = x) \tag{3.9}$$

$$= \sum_{x=0}^{\infty} \frac{x-1}{x} \times p_{\text{on}}^{W}(x) = 1 - E\left\{\frac{1}{T_{\text{on}}^{W}}\right\},$$

which can be easily calculated with known $p_{\text{on}}^{W}(x)$.

To obtain the EST of each content file, we first calculate the EST of one encoded packet. Based on the repeat-after-interruption mode in [16], let $s_{n,\ell}^{W}(x)$ denote the probability that the remaining EST of a packet with size n bits equals x slots given that the remaining service time is ℓ slots and that the slot preceding the remaining EST is an on-slot. Thus, we have $s_{n,\ell}^{W}(x) = 0$ for $x < \ell$. For $x \geq \ell$, the following equation can be derived by considering the on–off state of the first slot of the remaining EST:

$$s_{n,\ell}^{W}(x) = \delta s_{n,\ell-1}^{W}(x-1) + (1-\delta) \sum_{j=1}^{\infty} p_{\text{off}}^{W}(j) s_{n,n-1}^{W}(x-j-1). \tag{3.10}$$

Denote by $S_{n,\ell}^{W}(z)$ and $P_{\text{off}}^{W}(z)$ the probability generating functions (pgfs) of $s_{n,\ell}^{W}(x)$ and $p_{\text{off}}^{W}(x)$, respectively, i.e.,

$$P_{\text{off}}^{W}(z) = \sum_{x=0}^{\infty} p_{\text{off}}^{W}(x) z^{x}, \quad S_{n,\ell}^{W}(z) = \sum_{x=0}^{\infty} s_{n,\ell}^{W}(x) z^{x}. \tag{3.11}$$

Thus, we have the following lemma:

Lemma 3.1 *The average EST (in slots) of delivering a packet with size n bits through Wi-Fi transmission, denoted by $\overline{T}_{n}^{W}$, is*

$$\overline{T}_{n}^{W} = \frac{\delta}{1-\delta}\left(1 + (1-\delta)\mu_{\text{off}}^{W}\right)\left(\frac{1}{\delta^{n}} - 1\right). \tag{3.12}$$

Proof According to (3.10) and (3.11), the pgf of $s_{n,\ell}^{W}(x)$ is

$$S_{n,\ell}^{W}(z) = \sum_{x=1}^{\infty} \delta s_{n,\ell-1}^{W}(x-1) z^{x} + \sum_{x=1}^{\infty}(1-\delta) \sum_{j=1}^{\infty} p_{\text{off}}^{W}(j) s_{n,n-1}^{W}(x-j-1) z^{x}$$

$$= \delta z \sum_{x=0}^{\infty} s_{n,\ell-1}^{W}(x) z^{x} + (1-\delta)z \sum_{x=1}^{\infty} \sum_{j=1}^{\infty} p_{\text{off}}^{W}(j) z^{j} s_{n,n-1}^{W}(x-j-1) z^{x-j-1}$$

$$= \delta z S_{n,\ell-1}^{W}(z) + (1-\delta)z P_{\text{off}}^{W}(z) S_{n,n-1}^{W}(z).$$

Let $\zeta = (1-\delta)z P_{\text{off}}^{W}(z)$. Then the above equation can be simplified as

$$S_{n,\ell}^{W}(z) = \delta z S_{n,\ell-1}^{W}(z) + \zeta S_{n,n-1}^{W}(z).$$

By substituting different values for ℓ, we have

- $\ell = n$: $S_{n,n}^W(z) = \delta z S_{n,n-1}^W(z) + \zeta S_{n,n-1}^W(z) \Rightarrow S_{n,n}^W(z) = (\delta z + \zeta) S_{n,n-1}^W(z)$;
- $\ell = n-1$: $S_{n,n-1}^W(z) = \delta z S_{n,n-2}^W(z) + \zeta S_{n,n-1}^W(z) \Rightarrow S_{n,n-1}^W(z) = \frac{\delta z}{1-\zeta} S_{n,n-2}^W(z)$;
- $\ell = n-2$: $S_{n,n-2}^W(z) = \delta z S_{n,n-3}^W(z) + \zeta S_{n,n-1}^W(z) \Rightarrow S_{n,n-1}^W(z) = \frac{(\delta z)^2}{1-\zeta-\zeta\delta z} S_{n,n-3}^W(z)$;
- $\ell = n-3$: $S_{n,n-3}^W(z) = \delta z S_{n,n-4}^W(z) + \zeta S_{n,n-1}^W(z)$

$$\Rightarrow S_{n,n-1}^W(z) = \frac{(\delta z)^3}{1 - \zeta - \zeta\delta z - \zeta(\delta z)^2} S_{n,n-4}^W(z);$$

$\ldots$

By deductive proof, we have

$$S_{n,n-1}^W(z) = \frac{(\delta z)^{n-1}}{1 - \zeta \sum_{j=0}^{n-2}(\delta z)^j} S_{n,0}^W(z) = \frac{(\delta z)^{n-1}(1-\delta z)}{1 - \delta z - \zeta[1-(\delta z)^{n-1}]},$$

$$S_{n,n}^W(z) = (\delta z + \zeta) \frac{(\delta z)^{n-1}(1-\delta z)}{1 - \delta z - \zeta[1-(\delta z)^{n-1}]}$$

$$= \frac{\left(\delta z + (1-\delta)z P_{\text{off}}^W(z)\right)(\delta z)^{n-1}(1-\delta z)}{1 - \delta z - (1-\delta)z P_{\text{off}}^W(z)\left[1-(\delta z)^{n-1}\right]}.$$

Notice that if there is no more data to send, the downloading process ends in the current slot with probability 1, and thus we have $S_{n,0}^W(z) = 1$. Based on the pgf's moment generating property, the average EST $\overline{T}_n^W$ (in slots) of transmitting a packet with size n bits can be obtained by

$$\overline{T}_n^W = \left.\frac{d S_{n,n}^W(z)}{dz}\right|_{z=1} = \frac{\delta}{1-\delta}\left(1 + (1-\delta)\mu_{\text{off}}^W\right)\left(\frac{1}{\delta^n} - 1\right),$$

which concludes the proof. $\square$

Next, we analyze the delivery delay of content file f_m with K_m^W packets, each of which is of size α_m^W. After waiting for $\overline{T}_{wait}^W$ slots, the following slot is an on-slot which can serve one unit of data. Then, the EST of delivering the remaining $\alpha_m^W - 1$ units of the first packet can be obtained by replacing n by $\alpha_m^W - 1$ following Lemma 3.1. After the first packet transmission, the remaining $K_m^W - 1$ packets' delivery is preceded by an on-slot as the last slot of the previous packet's delivery is always an on-slot. Thus, each of the remaining $K_m^W - 1$ packets has an EST of $\overline{T}_{\alpha_m}^W$ slots. Thus, the average EST (i.e., delivery delay) of file f_m, denoted by $\overline{D}_m^W$, is expressed as

$$\overline{D}_m^W = (\overline{T}_{wait}^W + 1 + \overline{T}_{\alpha_m^W - 1}^W + (K_m^W - 1)\overline{T}_{\alpha_m^W}^W) \times l. \tag{3.13}$$

As shown in (3.13), the EST of a content file is determined by the on–off distribution, network capacity, and the content file size. The analysis procedure can be extended to the calculation of $\overline{D}_m^{TV}$, the average delay of downloading file f_m from TVWS stations, with different on–off distributions and network capacities.

Recall that content files are cached without coding in CBSs. When a file transmission is not completed within the range of one CBS, a CAI transmission mode can be applied when a vehicle travels through multiple CBSs. In this case, no retransmission is required and the average delivery delay of file f_m is

$$\overline{D}_m^C = z_m / \overline{R}_C. \tag{3.14}$$

For content files not cached in the HetVNet APs, they should be served by the CBS through backhaul links. Without loss of generality, CBSs are connected to the core network with wired backhaul links. Referring to [17], one hop can be assumed for wired backhauls. Thus, the average delay for retrieving file f_m with size z_m through wired backhaul links is

$$\overline{D}_m^B = \overline{D}_m^C + \left(\left(1 + 1.28\tfrac{\lambda_b}{\lambda_g}\right)\kappa\right) \cdot (a + bz_m), \tag{3.15}$$

where the second term is the backhaul transmission delay, λ_b and λ_g are the densities of CBSs and gateways, respectively. a, b, and κ are constants that reflect the processing capability of the nodes.

3.3 Matching-Based Content Caching Scheme

In this section, the content placement is optimized to minimize the overall content delivery delay based on the optimized coding parameters and the delivery delay analysis given in Sect. 3.2. Note that (3.3) is an ILP problem, which can be optimized by using ILP algorithms, e.g., branch and bound (B&B) algorithm. Considering the high complexity of the B&B algorithm, a more efficient solution is needed.

Construct a weighted bipartite graph $\mathcal{G} = (\mathcal{V}, \mathcal{I}, \mathcal{E})$, where $\mathcal{V}$ is the set of content files, $\mathcal{I}$ is the set of HetVNet APs, and $\mathcal{E}$ is the set of edges connecting vertices in $\mathcal{V}$ and $\mathcal{I}$. The weight of each edge can be defined based on the content delivery delay as analyzed in Sect. 3.2. Thus, the content placement problem is actually a weighted bipartite b-matching problem, which aims to find a subgraph $\mathcal{M} \subset \mathcal{G}$ to minimize the sum weight for the matched edges such that each vertex in $\mathcal{M}$ has at most b edges. Specifically, each file has $b = 1$, but the values of b for the APs are unknown for two reasons: (1) HetVNet APs have different storage capacities and (2) caching different files in the same AP requires different storage sizes. Therefore, b-matching algorithms are not suitable for the content placement optimization. Here we adopt the idea of the SA model and apply GS-based stable matching algorithm to match content files with HetVNet APs.

3.3.1 Complexity Analysis

As stated above, the content placement problem can be optimized by utilizing the B&B algorithm. Although the B&B can find the optimal solution, its worst-case time complexity is as high as that of brute-force exhaustive search. In this chapter, considering that each file has four caching possibilities (i.e., cached in Wi-Fi RSUs, TVWS stations, and CBSs or uncached), the worst-case complexity of the B&B algorithm is $O(4^M)$, where M is the total number of content files. Although the practical searching complexity is lower than 4^M in most cases, and the average complexity of the B&B algorithm can be reduced to be polynomial under some conditions, there is no complexity guarantee and the effectiveness of B&B is still limited by the potential exponential growth of the execution time as a function of problem size.

SA model solves the matching between students and colleges that have limited quotas, based on two-sided preferences. By adopting the SA model, we can map content files to be students and the HetVNet APs to be colleges. The content placement problem can then be formulated as a many-to-one matching problem between content files and the HetVNet APs. That is, one file can only be cached in one type of APs, while one type of APs can cache multiple files up to its quota (i.e., storage capacity). Then, the GS-based algorithm can be utilized to solve the matching problem with a much lower time complexity, which is $O(4 \times M)$.

3.3.2 Preference Lists

The matching between files and the caching APs is performed based on two-sided preferences, the construction of which can significantly affect the matching results and further the caching performance. Basically, the two-sided preference lists should be defined related to but not necessarily the same as the optimization objective. In this chapter, a multi-objective preference list construction is considered, by using two different metrics to define the preference lists for content files and the APs.

To minimize the overall delivery delay, the preferences of content files over HetVNet APs can be measured by the average delivery delay. Specifically, file f_m's preference over the APs is expressed as

$$\mathcal{P}_{\text{files}}(f_m, I) = \overline{D}_m^I, \tag{3.16}$$

where I refers to different methods of content delivery, i.e., Wi-Fi RSUs, TVWS stations, and CBS transmissions. In other words, $\mathcal{P}_{\text{files}}(f_m, \text{Wi-Fi}) = \overline{D}_m^W$, $\mathcal{P}_{\text{files}}(f_m, \text{TVWS}) = \overline{D}_m^T$, and $\mathcal{P}_{\text{files}}(f_m, \text{CBS}) = \overline{D}_m^C$. Basically, it is preferred that a content file is cached in the type of APs leading to the lowest delivery delay. Thus, by sorting the elements in $\mathcal{P}_{\text{files}}(f_m, I)$ in ascending order, the first type of APs in f_m's preference list is the most preferred APs to cache f_m.

Notice that content delivery delay is largely affected by file size. Thus, if using the delivery delay to define APs' preference lists, all the APs will prefer to cache small files, regardless of the file popularity. To address this issue, we define a new metric to rank the files based on file popularity and the amount of data that can be off-loaded from the backhaul links. In other words, APs prefer to cache files that have a higher request probability with a larger requested data size. In this way, APs can leverage their storage capacities more efficiently and off-load more traffic with lower delivery latency. Thus, the APs' preferences over file f_m are measured by

$$\mathcal{P}_I(I, f_m) = p_{\text{req}}^m \cdot \alpha_m^I \cdot K_m^I, \tag{3.17}$$

where the definition of I is the same as in (3.16). By sorting the elements in $\mathcal{P}_I(I, f_m)$ in descending order, the first file in each type of APs' preference list is file f_m leading to the maximum average offloading data size.

3.3.3 Matching-Based Content Placement Policy

The matching-based HetVNet content placement scheme with on–off service model is demonstrated in Algorithm 2. Firstly, for each content file, the average delivery delay is analyzed for all HetVNet APs, based on which the preference lists are constructed as discussed in (3.16) and (3.17). After that, the GS-based algorithm is exploited to solve the SA-based many-to-one matching problem between content files and APs. The matching process is described as follows:

Step 1: Each content file proposes to its current most favorite caching APs and then removes this type of APs from its preference list.

Step 2: Each type of APs check all the received proposals from the files, including the new proposals and those accepted in previous iterations, and then accept the most preferred files within the storage capacity constraint and reject the rest.

Step 3: For the rejected files, go to **Step 1**. The matching process terminates when all files are successfully cached or all APs' storage capacities are occupied.

Algorithm 2: Matching-based caching optimization algorithm

$\mathcal{F}, \mathcal{F}_u$: Sets of all the content files and unmatched content files, respectively.
z_m: Size of content file f_m. $\mathcal{P}_{\text{files}}(f_m, I)$: Preference lists for content files.
$\mathcal{P}_I(I, f_m)$: Preference lists for HetVNet APs.
C_T, C_W, C_C: Storage capacities of Wi-Fi RSUs, TVWS stations, and CBSs.
a_m^W, a_m^T, a_m^C: Indicators for caching file f_m in Wi-Fi RSUs, TVWS stations, and CBS.
begin
    Initialize $\mathcal{F}_u = \mathcal{F}$.
    repeat
        for $f_m \in \mathcal{F}_u$ **do**
            Propose to the first type of APs I in its preference list $\mathcal{P}_{\text{files}}(f_m, I)$.
            Set $a_m^I = 1$ ($a_m^I \in \{a_m^W, a_m^T, a_m^C\}$) and remove I from $\mathcal{P}_{\text{files}}(f_m, I)$.
        end
        for $I \in$ *Wi-Fi RSUs, TVWS stations, CBSs* **do**
            $S_{I,\text{req}}^m = z_m$ if I is CBS; otherwise, $S_{I,\text{req}}^m = \alpha_m^I \cdot n_m^I$.
            if $\sum_{m \in \mathcal{F}}(a_m^I \cdot S_{I,\text{req}}^m) \leq C_I$ **then**
                I keeps all the proposing files and removes accepted files from $\mathcal{F}_u$.
            else
                I keeps the most preferred files under storage capacity constraint and rejects the rest;
                Remove these accepted files from $\mathcal{F}_u$.
                For the rejected files, set $a_m^I = 0$ and add them into $\mathcal{F}_u$.
            end
            $C_{I,remain} = C_I - \sum_{m \in \mathcal{F}}(a_m^I \cdot S_{I,\text{req}}^m)$
        end
    until $\mathcal{F}_u = \emptyset$ *or* $C_{I,remain} \leq \min\limits_{f_m \in \mathcal{F}_u} S_{I,\text{req}}^m, \forall I$;
    Output: a_m^W, a_m^T, and a_m^C for any $f_m \in \mathcal{F}$.
end

3.4 Performance Evaluation

We conduct simulations based on the real scenario of University of Waterloo campus. The campus map is shown in Fig. 3.3a and the main roads are depicted in Fig. 3.3b. Vehicles in the target region can always access the CBS, which is marked as the pink triangle in Fig. 3.3b. Blue circles in Fig. 3.3b represent the coverage areas of five Wi-Fi RSUs, and the red circles are the coverage areas of three TVWS stations. To simulate realistic vehicle traffic, we use VISSIM simulation tool to generate the traffic of 200 vehicles in the campus scenario. The content file size is within the range of [0 MB, 1000 MB]. The file popularity follows Zipf distribution with exponent $\xi = 0.7$. The default values of main simulation parameters are listed in Table 3.2 unless otherwise specified.

Recall that we assume known distributions of the on–off periods in this chapter. In the simulation, based on the mobility traces generated by VISSIM and the AP deployment, we can easily calculate the time duration vehicles spend in each AP, and thus the on–off period distributions can be acquired. For instance, in our simulation, after processing the vehicle trace data, the on–off distributions for the TVWS

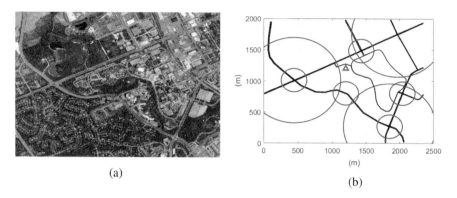

(a)

(b)

Fig. 3.3 Simulation settings. (**a**) Simulation scenario map. (**b**) AP deployment

Table 3.2 Simulation parameters

$[r_W, r_T]$: coverage radii of Wi-Fi RSUs and TVWS stations	[150, 600] m
$[R_W^a, R_T^a, R_C^a]$: aggregate rates of a Wi-Fi RSU, TVWS station, and CBS	[65, 54, 128] Mbps
c and ϵ: constant in Robust Soliton Distribution and decoding failure probability	0.1 and 0.05
p_{suc}^W and p_{suc}^T: probabilities that vehicles can download at least one encoded packet from Wi-Fi RSUs and TVWS stations	0.99
p_{max}^W and p_{max}^T: probabilities that vehicles can download enough packets from Wi-Fi RSUs and TVWS stations without waiting	0.9
$[C_W, C_T, C_C]$: storage capacity of each Wi-Fi RSU, TVWS station, and CBS	[10, 10, 20] GB

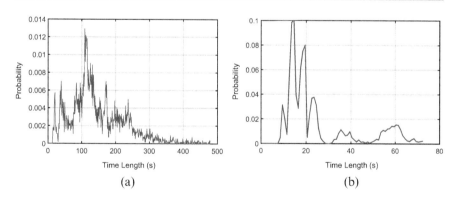

(a)

(b)

Fig. 3.4 Distributions of on–off periods for TVWS transmission. (**a**) On-periods. (**b**) Off-periods

transmission can be obtained and shown in Fig. 3.4. Note that the distributions vary in different target regions with different road layout and AP deployment. Simulations in this section leverage the on–off TVWS service time distributions in Fig. 3.4, and the on–off Wi-Fi service time distribution can be obtained in a similar

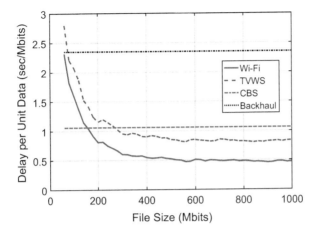

Fig. 3.5 Average delay per unit data vs. file size

way. However, the proposed caching scheme can be applied to scenarios with any other well-known distributions for the on–off periods.

To evaluate the delay performance, we monitor all vehicles in the target region, each of which generates content requests based on the file popularity distribution. Then a vehicle is randomly chosen at a random time instant to observe its data downloading performance. All the simulation results are averaged over 1000 trials.

Impact of File Size Figure 3.5 shows the impact of file size on the average delay performance[3] of content delivery from Wi-Fi RSUs, TVWS stations, CBS, or backhaul transmission. Since the average transmission rates of the CBS and backhaul delivery are mainly determined by the number of vehicles sharing the spectrum and the deployment of CBSs, the average delays of CBS and backhaul transmissions keep unchanged with increasing file size, as shown in Fig. 3.5. The average delay for Wi-Fi and TVWS transmissions declines with increasing file size, and the former has a better delay performance than the latter. For small-size files, Wi-Fi and TVWS transmissions have poor delay performance. The reason is that, when compared with the average waiting time, the required service time for small files plays a minor part in the EST in our on–off service models, leading to a large delay per unit data. With increasing file size, a larger proportion of the EST comes from the service time instead of the waiting time. Finally, the average delay per unit data converges to a constant value which is determined by the average Wi-Fi/TVWS transmission rate and the ratio of the average time length of on-periods over that of off-periods. Therefore, it is advisable that small files (e.g., texts or pictures) are

---

[3] Since overall content delivery delay is significantly affected by file size, average delay per unit data (sec/bit), which is the ratio of the overall delay defined in (3.3) over the total requested file size, is adopted in the simulation to represent the content delivery delay performance.

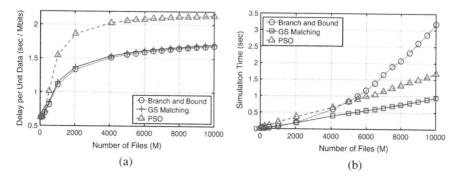

Fig. 3.6 Delay and complexity performance comparison between B&B, PSO, and GS matching-based schemes. (**a**) Average delay per unit data. (**b**) Time complexity comparison

cached in the CBS without coding, while large files should be encoded and cached in Wi-Fi RSUs or TVWS stations to reduce the delivery delay.

Trade-off Between Delay Performance and Complexity Figure 3.6 shows the impact of the number of content files on the achievable delay performance and complexity of different schemes. In addition to the B&B and the proposed matching-based schemes, one evolutionary scheme, the particle swarm optimization (PSO)-based scheme, is also included for performance comparison. Intuitively, with more content files in the network, the delivery delay per unit data and the time complexity for all the schemes increase. As shown in Fig. 3.6a, the delivery delay increases because more files need to be retrieved via backhaul links because of the limited storage capacity. The B&B-based scheme outperforms the proposed GS-based matching scheme, while the performance gap is insignificant. On the other hand, compared to the B&B-based scheme which has exponentially increased complexity over the network size, the GS-based matching scheme is significantly less time-consuming, as shown in Fig. 3.6b, especially when the system scales. The PSO-based scheme achieves a larger delivery delay with a longer simulation time when compared with the matching-based scheme. The delay performance of the PSO-based scheme can be further improved, while the corresponding simulation time will also increase significantly. Therefore, the matching-based scheme is a favorable choice to reduce the time complexity with modest delay performance loss, especially in complex or heterogeneous networks with a large number of files.

Coded Caching vs. Uncoded Caching Different from the coded caching adopted in this chapter, uncoded caching means that files are cached entirely in APs. The CAI transmission mode can be applied when the vehicle travels through multiple APs. With uncoded caching and CAI transmission mode, one may naively believe that the delivery delay can be reduced since no retransmission is required. However, uncoded caching leads to a low storage efficiency, which further affects the overall delay and offloading performance. Therefore, in this part, we compare performances of the coded caching and uncoded caching schemes to dispel any wishful thinking.

Similar to the analysis in Sect. 3.2.3, the EST of transmitting a file with size z_m by using the CAI transmission mode is analyzed as follows (taking Wi-Fi transmission as an example). Given that the preceding slot is an on-slot, the probability that the EST of a packet of length n bits equals ℓ slots is

$$s_n^{W,\text{CAI}}(\ell) = \delta s_{n-1}^{W,\text{CAI}}(\ell - 1) + (1 - \delta) \sum_{j=1}^{\infty} P_{\text{off}}^W(j) s_{n-1}^{W,\text{CAI}}(\ell - j - 1). \tag{3.18}$$

The corresponding pgf of $s_n^{W,\text{CAI}}(\ell)$ is

$$S_n^{\text{CAI}}(z) = \left[\delta z + (1 - \delta) z P_{\text{off}}^W(z) \right] S_{n-1}^{\text{CAI}}(z) = \left[\delta z + (1 - \delta) z P_{\text{off}}^W(z) \right]^n. \tag{3.19}$$

Thus, the EST of transmitting a file with size z_m in CAI mode is

$$\overline{D}_m^{W,\text{CAI}} = \left(\overline{T}_{wait}^{W,\text{CAI}} + 1 + \frac{d S_{\alpha_m - 1}^{\text{CAI}}(z)}{dz} \bigg|_{z=1} \right) \times l \tag{3.20}$$

$$= \left(\frac{(1 - \sigma) \mu_{\text{off}}^W}{2} + 1 + (z_m - 1) \left[1 + (1 - \delta) \mu_{\text{off}}^W \right] \right) \times l.$$

Figure 3.7 shows the delay and offloading performances of coded and uncoded content caching schemes. As shown in Fig. 3.7a, the average delay per unit data for both caching schemes increases with more content files as more files need to be fetched via backhaul links. It is worth noting that when the number of files is larger than 40, coded caching scheme outperforms the uncoded scheme in terms of

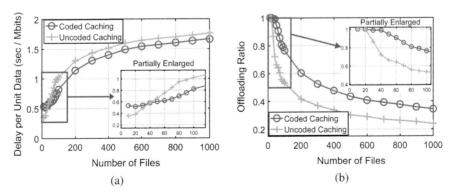

Fig. 3.7 Delay and offloading performance comparison for coded caching and uncoded caching schemes. (**a**) Average delay per unit data. (**b**) Offloading performance

the average delay, since the former can cache more content files in the APs due to higher storage efficiency.

Define offloading ratio as the ratio of the data volume downloaded without going through backhaul links over the overall requested data volume. As shown in Fig. 3.7b, the offloading ratios of the coded and uncoded caching schemes are identical when there are less than 20 files, since the HetVNet APs can successfully cache all the files. With increasing number of files, a smaller portion of files can be cached in the APs for uncoded caching scheme, leading to higher probability of backhaul transmission and lower offloading ratio. On the other hand, despite the decline of offloading ratio, the coded caching scheme has significant advantage in terms of offloading performance when compared with the uncoded scheme. Stemming from the above observations, uncoded caching is preferred in scenarios with a small number of content files, while coded caching scheme performs better when network scales.

Single-Access-Based Caching vs. Multi-access-Based Caching In the proposed caching scheme, a file is allowed to be cached in only one type of APs to improve caching storage efficiency without requiring cooperation among different access networks. Intuitively, for the cached files, the delivery delay can be further reduced if they are cached in all the APs such that vehicles can get served within any access network coverage. However, whether the overall delay performance can be improved remains unknown. In this part, we compare the proposed caching scheme with the case where files are encoded and cached in all the APs, named as "Multi-Access Caching," to reveal insights on the suitability of these two types of caching schemes in different scenarios. Notice that, in "Multi-Access Caching" scheme, files are encoded with the same coding parameters [i.e., in Algorithm 1, let $\alpha_m = \min\{\alpha_m^W, \alpha_m^T\}$ and $k_m = z_m/\alpha_m$, then n_m^I and K_m^I (I refers to Wi-Fi RSUs, TVWS stations, or CBSs) can be calculated accordingly].

In Fig. 3.8, we compare the performance of the proposed scheme and the "Multi-Access Caching" scheme with different caching storage ratios.[4] With increasing storage capacities, both caching schemes achieve a higher cache hit rate, lower delivery delay, and better offloading performances. As shown in Fig. 3.8a, c, the proposed scheme can achieve a high cache hit rate and offloading ratio since more content files can be cached in the HetVNet APs for backhaul offloading. Although the cached files can be delivered with a lower delay in the "Multi-Access Caching" scheme, the overall delivery delay performance is unsatisfactory due to the substantial backhaul transmission when the caching storage capacity is limited. Nevertheless, the "Multi-Access Caching" scheme outperforms the proposed scheme in terms of overall delivery delay when the storage capacity is large enough, e.g., when the caching storage of each AP is no less than 0.9 of the total size of all the files as shown in Fig. 3.8b. To summarize, the "Multi-Access

---

[4] All the APs are assumed to have the same storage capacity, and the caching storage ratio is the ratio of the storage capacity of one AP over the total size of all the content files.

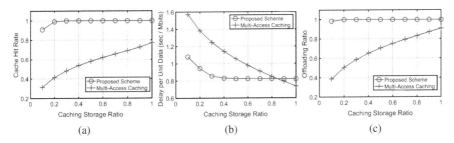

Fig. 3.8 Cache hit rate, delay, and offloading performance comparison between the proposed scheme and multi-access-based caching scheme. (**a**) Cache hit rate. (**b**) Average delay per unit data. (**c**) Offloading performance

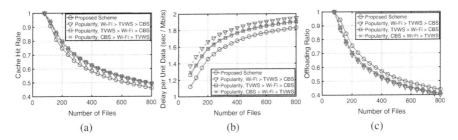

Fig. 3.9 Cache hit rate, delay, and offloading performance comparison between the proposed scheme and popularity-based caching schemes. (**a**) Cache hit rate. (**b**) Average delay per unit data. (**c**) Offloading performance

Caching" scheme is preferred when the storage capacity is sufficiently large, while the proposed scheme works well in general cases with limited caching resources.

Performance Comparison with Popularity-Based Caching Schemes The proposed caching scheme is also compared with the popularity-based caching schemes. In particular, the popularity-based schemes prioritize and cache the files based only on file popularity. As shown in Fig. 3.9, we use "Popularity" to denote the popularity-based caching schemes. In addition, "Wi-Fi > TVWS > CBS" denotes that content files are first cached in Wi-Fi RSUs until reaching the caching capacity and then in the TVWS stations, and the CBSs have the lowest priority. "TVWS > Wi-Fi > CBS" and "CBS > Wi-Fi > TVWS" are defined in a similar way. In addition to the average delay and offloading performance, the cache hit rate[5] is also considered to further compare the performance of the proposed scheme and popularity-based schemes.

As shown in Fig. 3.9, with a small number of content files, all content files can be cached in the APs, thus the cache hit rate and offloading ratio are both equal to 1

---

[5] The cache hit rate is defined as the ratio of the number of cache hit in all the APs' caches to the overall number of vehicular content requests.

for all the schemes. With increasing numbers of files, more files need to be retrieved from backhaul links, thereby leading to a lower cache hit rate, longer delivery delay, and lower offloading ratio for all the schemes. As shown in Fig. 3.9a, the popularity-based schemes have a higher cache hit rate than the proposed scheme since the former ones only cache the most popular files. On the other hand, in addition to the file popularity, the proposed scheme also takes file size, vehicle mobility, and network characteristics into consideration to ably cache different types of files in the APs. Therefore, the proposed scheme presents better delay and offloading performances as shown in Fig. 3.9b, c.

3.5 Summary

In this chapter, we have investigated encoded content caching in terrestrial Het-VNets to provide enhanced and diversified wireless network access for moving vehicles and reduce content delivery delay, by considering the impact of factors including file popularity, vehicle mobility, network service interruption, and storage capacities of the APs. Specifically, we have proposed a matching-based scheme with multi-objective two-sided preference lists to optimize the content placement problem. Simulation results have validated the effectiveness of the proposed content caching scheme, which can further provide an insight into the optimization of content sharing in different network conditions.

References

1. H. Wu, J. Chen, C. Zhou, W. Shi, N. Cheng, W. Xu, W. Zhuang, X. Shen, Resource management in space-air-ground integrated vehicular networks: SDN control and AI algorithm design. IEEE Wirel. Commun. 27(6), 52–60 (2020)
2. N. Wang, J. Wu, Opportunistic WiFi offloading in a vehicular environment: waiting or downloading now? in Proc. IEEE INFOCOM 2016, San Francisco, CA, April 2016
3. J.-H. Lim, K. Naito, J.-H. Yun, M. Gerla, Reliable safety message dissemination in NLOS intersections using TV white spectrum. IEEE Trans. Mobile Comput. 17(1), 169–182 (2018)
4. Y. Han, E. Ekici, H. Kremo, O. Altintas, Vehicular networking in the TV white space band: challenges, opportunities, and a media access control layer of access issues. IEEE Veh. Technol. Mag. 12(2), 52–59 (2017)
5. W. Wu, N. Zhang, N. Cheng, Y. Tang, K. Aldubaikhy, X. Shen, Beef up mmWave dense cellular networks with D2D-assisted cooperative edge caching. IEEE Trans. Veh. Technol. 68(4), 3890–3904 (2019)
6. J. Chen, H. Wu, P. Yang, F. Lyu, X. Shen, Cooperative edge caching with location-based and popular contents for vehicular networks. IEEE Trans. Veh. Technol. 69(9), 10291–10305 (2020)
7. L. Wang, H. Yang, X. Qi, J. Xu, K. Wu, iCast: fine-grained wireless video streaming over internet of intelligent vehicles. IEEE Internet Things J. 6(1), 111–123 (2019)
8. Y. Lin, B. Liang, B. Li, Data persistence in large-scale sensor networks with decentralized fountain codes, in Proc. IEEE INFOCOM 2007, Barcelona, May 2007

9. L. Wang, H. Wu, Y. Ding, W. Chen, H.V. Poor, Hypergraph based wireless distributed storage optimization for cellular D2D underlays. IEEE J. Sel. Areas Commun. **34**(10), 2650–2666 (2016)

10. D. Gale, L.S. Shapley, College admissions and the stability of marriage. Am. Math. Mon. **69**(1), 9–15 (1962)

11. O. Kaiwartya, A.H. Abdullah, Y. Cao, A. Altameem, M. Prasad, C.-T. Lin, X. Liu, Internet of vehicles: motivation, layered architecture, network model, challenges, and future aspects. IEEE Access **4**, 5356–5373 (2016)

12. N. Garg, M. Sellathurai, V. Bhatia, B. Bharath, T. Ratnarajah, Online content popularity prediction and learning in wireless edge caching. IEEE Trans. Commun. **68**(2), 1087–1100 (2020)

13. H. Wu, F. Lyu, C. Zhou, J. Chen, L. Wang, X. Shen, Optimal UAV caching and trajectory in aerial-assisted vehicular networks: a learning-based approach. IEEE J. Sel. Areas Commun. **38**(12), 2783–2797 (2020)

14. M. Jain, G.C. Sharma, R. Sharma, Maximum entropy approach for discrete-time unreliable server $Geo^X/Geo/1$ queue with working vacation. Int. J. Math. Oper. Res. **4**(1), 56–77 (2012)

15. M. Luby, LT codes, in *Proc. IEEE FOCS 2002, Vancouver, BC*, Nov 2002, pp. 271–280

16. D. Fiems, B. Steyaert, H. Bruneel, Discrete-time queues with generally distributed service times and renewal-type server interruptions. Perform. Eval. **55**(3–4), 277–298 (2004)

17. G. Zhang, T.Q.S. Quek, M. Kountouris, A. Huang, H. Shan, Fundamentals of heterogeneous backhaul design—analysis and optimization. IEEE Trans. Commun. **64**(2), 876–889 (2016)

Chapter 4
Optimal UAV Caching and Trajectory Design in the AGVN

Abstract In this chapter, we investigate the UAV-assisted mobile edge caching to assist terrestrial vehicular networks in delivering high-bandwidth content files. To maximize the overall network throughput, we formulate a joint caching and trajectory optimization (JCTO) problem to jointly optimize content placement, content delivery, and UAV trajectory. Considering the intercoupled decisions and the limited UAV energy, the formulated JCTO problem is intractable directly and timely. Therefore, we propose a deep supervised learning (DSL) scheme to enable intelligent edge for real-time decision-making in the highly dynamic vehicular networks. Specifically, we first propose a clustering-based two-layered (CBTL) algorithm to solve the JCTO problem offline. With a given content placement strategy, we devise a time-based graph decomposition method to jointly optimize the content delivery and trajectory design, with which we then leverage the particle swarm optimization (PSO) algorithm to further optimize the content placement. We then design a convolutional neural network (CNN)-based DSL architecture to make fast decisions online. The network density and content request distribution with spatial–temporal dimensions are labeled as channeled images and input to the CNN-based model, and the results achieved by the CBTL algorithm are labeled as model outputs. With the CNN-based model, a function mapping the input network information to output decisions can be intelligently learnt to make timely decisions. Extensive trace-driven experiments are carried out to demonstrate the efficiency of CBTL in solving the JCTO problem and the superior learning performance with the CNN-based model.

To accommodate ever-increasing CAV traffic demands especially in the future driverless era, the CAV networks are envisioned to be vigorously robust for delivering high-bandwidth content files [1, 2]. As the VNs are highly dynamic, the fixed and rigid terrestrial network resources can hardly guarantee satisfactory network performance anywhere at any time, especially at urban busy roads during rush hours. UAV-assisted mobile edge caching is a promising paradigm to assist terrestrial networks. By proactively caching popular and repetitively requested content files with large size (such as HD map and the video streaming of a football

© The Author(s), under exclusive license to Springer Nature Switzerland AG 2022   61
H. Wu et al., *Mobile Edge Caching in Heterogeneous Vehicular Networks*,
SpringerBriefs in Computer Science, https://doi.org/10.1007/978-3-030-88878-7_4

match or a concert), edge caching can significantly alleviate the terrestrial network burden [3, 4]. In addition, caching-enabled UAVs are physically free from backhaul limitations, which makes the UAV implementation more feasible considering the high agility of UAVs. With fully controllable mobility and high altitude, UAVs can provide high LoS probability for UAV-to-vehicle (U2V) wireless links. Besides, UAVs can be dispatched on demand when the terrestrial network is overloaded, the manner of which is flexible and cost-effective.

Despite the existing works mentioned in Chap. 2, there are still various technical challenges associated with UAV-assisted mobile edge caching in VNs. First, most existing works study UAV caching in scenarios with low/no user mobility, which cannot be directly applied to the VNs. The spatial–temporal variation in vehicle density and content request distribution affects the UAV caching performance, which should be considered in the caching scheme design. Second, the joint decision of content caching, UAV trajectory, and content delivery in the AGVN has not been well addressed. The joint optimization is essential to improve the UAV content delivery performance as the three decisions interact with one another. Third, as the vehicular network condition varies significantly, online decisions are required to keep pace with the dynamic vehicular environments, posing real-time requirements to the optimization solution.

In this chapter, we focus on the joint design of UAV caching (including content placement and delivery) and UAV trajectory in urban vehicular networks that have substantial content demands. The objective is to find the optimal solution to the joint optimization problem in real time to maximize the overall network throughput under the UAVs' energy constraints. By partitioning the target area into small regular grids and representing each grid by a point, we construct a topology graph to find the optimal flying path for each UAV, where the edge weights are affected by the content placement and delivery scheme. As the formulated JCTO problem is intractable directly and timely due to the intercoupled variables, we propose a learning-based scheme named *LB-JCTO* to enable edge intelligence (EI) and make real-time decisions in the highly dynamic AGVNs.

LB-JCTO is an offline optimization and learning for online decision framework, in which the DSL architecture is leveraged to make online decisions under the supervision of offline optimized targets. Specifically, in the first stage of *LB-JCTO*, we propose a CBTL algorithm to solve the JCTO problem. Given a content placement strategy, we jointly optimize UAV trajectory and content delivery by a time-based graph decomposition method, where a directed graph is constructed by considering the spatial–temporal variant vehicle densities and content requests. A resource constrained shortest path (RCSP) algorithm [5] is then used to find the optimal path, representing the optimal content delivery and trajectory solution. Then we leverage the PSO algorithm for content placement optimization to further enhance the network throughput. Although the CBTL algorithm can achieve a satisfactory performance, the algorithm complexity cannot meet the real-time requirements for online decisions. Therefore, in the second stage of *LB-JCTO*, we adopt a CNN-based DSL framework [6] to conduct supervised learning at the intelligent edge. Particularly, the spatial–temporal variant network density and content requests are labeled as channeled images and input to the learning model, and the optimized solutions obtained by the CBTL algorithm are labeled as model outputs. With the well-trained CNN model, a function mapping the input network information to output decisions can be learnt to enable timely decision-making.

The merits of *LB-JCTO* are threefold: (1) at the offline optimization stage, we can use a complicated algorithm with a relatively high computation complexity to obtain optimized results, which are good learning targets, (2) for offline training, as EI-based *LB-JCTO* is trained to learn from a good target, its performance can be well guaranteed, and (3) at the online stage, with the well-trained CNN-based model, *LB-JCTO* simply runs a matching function in the UAVs for the up-to-date collected information, which can output high-quality and real-time decisions. We highlight our major contributions as follows:

- We study the joint design of UAV caching and trajectory in AGVNs, which is of significant importance for future CAVs to deliver high-bandwidth content files robustly. Specifically, we formulate the JCTO problem to investigate the interplay between the caching scheme design and UAV trajectory optimization, where the analysis and derivation of UAV energy consumption and achievable network throughput are, respectively, presented.
- To solve the JCTO problem in real time, we propose a learning-based scheme named *LB-JCTO* to make online decisions to respond to the dynamic vehicular networks. Specifically, in the *LB-JCTO* scheme, the CBTL algorithm is devised to optimize the JCTO problem offline, and then a CNN-based learning scheme is designed to enable online decisions.
- Extensive trace-driven experiments are conducted to demonstrate that the CBTL algorithm can efficiently solve the JCTO problem, and the CNN-based model has superior learning performance while satisfying the real-time requirements.

The remainder of this chapter is organized as follows. We present the system model and problem formulation in Sect. 4.1. Section 4.2 shows the proposed *LB-JCTO* scheme. Section 4.3 elaborates the CBTL-based offline optimization, and the CNN-based learning model is given in Sect. 4.4. Trace-driven experimental results are provided in Sect. 4.5, followed by a brief summary in Section 4.6.

4.1 System Scenario and Problem Formulation

4.1.1 Scenario Description

In this chapter, we investigate UAV-assisted mobile edge caching in the AGVN by focusing on a rectangle area covered by a single CBS, as shown in Fig. 4.1. A set $\mathcal{K}$ of K rotary-wing UAVs with caching capabilities is dispatched to serve the ground vehicles along with the CBS. Let $\mathcal{F} = \{f_1, f_2, \cdots, f_F\}$ be the set of F files

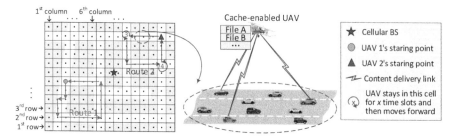

Fig. 4.1 Overview of UAV-aided edge caching in vehicular networks

requested by the vehicles. The size of content files is assumed to be the same[1] and denoted by $\varsigma_f, \forall f_f \in \mathcal{F}$. Denote by C_k the caching storage capacity of UAV k, i.e., UAV k can cache no more than C_k content files. In addition, UAVs are assumed to fly horizontally at a constant altitude of H to achieve a lower level of energy consumption [7].

The VNs are highly dynamic with spatial–temporal variant vehicle densities and content request distributions. Constrained by the street layout, vehicle densities on the roads (white areas in Fig. 4.1) are generally larger than those on the other areas, so as the content request probabilities. Without loss of generality, vehicle density can be estimated based on position information collected from GPS devices. Content popularity is modeled by the Zipf distribution according to the analysis for many real datasets [8]. With Zipf-based content popularity, the request distributions still vary spatially and temporally due to different user preferences. Inspired by the model in [9], we model the file preference in each grid at time slot t as follows:

- A grid v and a content file f_f are, respectively, associated with feature values $\phi_{v,t}$ and ϑ_f, where $\phi_{v,t}, \vartheta_f \in [0, 1]$.[2]
- The probability that file f_f is requested by users in grid v at time t is

$$r_{v,t,f} = p_f \frac{g(\phi_{v,t}, \vartheta_f)}{\sum_{v' \in \mathcal{V}} g(\phi_{v',t}, \vartheta_f)}, \tag{4.1}$$

where $p_f = \frac{1/f^\xi}{\sum_{m \in \mathcal{F}} 1/m^\xi}$ and $g(\phi_{v,t}, \vartheta_f) = (1 - |\phi_{v,t} - \vartheta_f|)^{\frac{1}{\sigma^3}-1}$ are the popularity of file f_f and the correlation function between grid v and file f_f.

---

[1] In reality, for files with different sizes, the analysis can be easily extended by dividing each file into chunks of equal size.

[2] Features of a location may include weather characteristics, historical park, and upcoming famous events; a content file is associated with features such as file type and size, metadata, keywords, or tags. To simplify the model, we use a random value to represent the features of a grid/file, but this basic model can be easily extended to multi-dimensional feature vectors.

Basically, $r_{v_1,t_1,f} \neq r_{v_2,t_2,f}$ for $v_1 \neq v_2$ or $t_1 \neq t_2$ due to its spatial–temporal variance, despite that vehicles may have similar preferences over some popular content files.

To maximize the overall network throughput, UAVs should move to different locations in response to the network dynamics. To better illustrate the time-varying locations of UAVs, we partition the target area into small square grids with side length w. Each grid square is represented by its central point and denoted by (i, j), $i \in [1, N_{\text{row}}]$, $j \in [1, N_{\text{col}}]$. This means that the grid is located in the i-th row and j-th column, and N_{row} and N_{col} are the numbers of rows and columns in the target area. Then a topology graph $G = (\mathcal{V}, \mathcal{E})$ can be constructed to represent the whole map, where $\mathcal{V}$ is the set of central points of the grid squares and $\mathcal{E}$ contains the edges connecting the central points. For two points $u = (i, j)$ and $v = (m, n)$, $(u, v) \in \mathcal{E}$ if and only if $u, v \in \mathcal{V}$, $|i - m| \leq 1$, $|j - n| \leq 1$.[3]

We consider a discrete-time system where the UAV endurance is equally discretized into T_U time slots, each with a duration of Δ_t. UAVs can fly along the points and edges in graph G. Notice that the starting and ending points (the location where the UAV is dispatched and collected for battery charging) of a trajectory are the same, while the starting/ending points vary for different UAVs. Figure 4.1 shows two UAV trajectories (marked in red and green) with different starting points. Along the flying trajectory, UAVs can keep changing locations (Route 1) or hover above a certain location for multiple time slots (Route 2).

To facilitate analysis, we define necessary notation and symbols below:

1. $\mathbf{D}_{N_{\text{row}} \times N_{\text{col}} \times T_U}$ is a three-dimensional array representing the vehicle densities. The (i, j, t)-th entry $d_{i,j,t}$ (or $d_{v,t}$ if $v = (i, j)$) is the average number of vehicles in the grid represented by point (i, j) $((i, j) \in \mathcal{V})$ at time slot t $(t \leq T_u)$.
2. $\mathbf{R}_{N_{\text{row}} \times N_{\text{col}} \times T_U \times F}$ is a four-dimensional array representing the content request distributions. The (i, j, t, f)-th entry $r_{i,j,t,f}$ (or $r_{v,t,f}$ if $v = (i, j)$) is the request probability of file f_f in grid (i, j) at time slot t.
3. $\mathbf{X}_{K \times T_U} = [\mathbf{x}_1, \mathbf{x}_2, \cdots, \mathbf{x}_K]$ is the flying trajectories of K UAVs. $\mathbf{x}_k = [x_{k,1}, x_{k,2}, \cdots, x_{T_U}]$ is the k-th UAV's trajectory within its endurance time, where $x_{k,t}$ is its location at time slot t. Thus we have $(x_{k,t}, x_{k,t+1}) \in \mathcal{E}$ and $x_{k,1} = x_{k,T_U} = v_{0,k}$, where $v_{0,k}$ is the starting point where UAV k is released.
4. $\mathbf{A}_{K \times F} = [\mathbf{a}_1, \mathbf{a}_2, \cdots, \mathbf{a}_K]$ is an indicator matrix showing the caching of content files in the UAVs, where $\mathbf{a}_k = [a_{k,1}, a_{k,2}, \cdots a_{k,F}]$ represents the k-th UAV's caching status. $a_{k,f} = 1$ if content file f_f is cached in UAV k; otherwise, $a_{k,f} = 0$.
5. $\mathbf{S}_{K \times T_U} = [\mathbf{s}_1, \mathbf{s}_2, \cdots, \mathbf{s}_K]$ denotes the content delivery decisions of UAVs along the trajectories, where $\mathbf{s}_k = [s_{k,1}, s_{k,2}, \cdots s_{k,T_U}]$. $s_{k,t} = 1$ if UAV k determines to serve vehicular requests at time t, whereas $s_{k,t} = 0$ means that UAV k flies without content delivery at time t.

---

[3] In the remainder of this chapter, v and (i, j) are used interchangeably to represent a grid square.

4.1.2 Communication and UAV Energy Consumption Models

UAV-to-Vehicle (U2V) Communications In this chapter, the mobile edge caching-enabled UAVs work in the Wi-Fi spectrum with a constant transmission power, which is denoted by P_U. Notice that NLoS links can severely degrade the communication performance and cannot support efficient U2V content delivery. Therefore, the UAVs' coverage radius, denoted by r_{UAV}, can be defined by constraining the LoS probability and free space pathloss [10]. With a fixed flying altitude H, we have

$$r_{\text{UAV}} = \min \left\{ \frac{H}{\tan\left(a_1 - \frac{1}{a_2} \ln\left(\frac{1-\xi_{LoS}}{a_1 \xi_{LoS}}\right)\right)}, \sqrt{\left(\frac{c\gamma_{\max}}{4\pi f_c}\right)^2 - H^2} \right\}, \tag{4.2}$$

where a_1 and a_2 are constant values determined by environment. f_c, c, ξ_{LoS}, and $\gamma_{\max}$ represent the carrier frequency, light speed, the LoS probability requirement, and U2V free space pathloss threshold, respectively. When partitioning the target area into grid squares, we set the side length of each square to be $w = \sqrt{2}r_{UAV}$. With this setting, when a UAV hovers above a grid square, its coverage area can approximately equal the square area.

According to IEEE 802.11 standard, the Wi-Fi coverage area can be divided into zone areas based on the achieved signal-to-noise ratio (SNR) levels, and the data rates are calculated based on the Wi-Fi modulation and coding schemes. Therefore, the coverage area of a Wi-Fi-based UAV can be divided into L zones, where the j-th zone has a distinct annulus area with width l_j and a data rate of R_j. Therefore, a vehicle within a grid area can achieve a mean throughput of

$$\overline{R}_{UAV} = \rho \left(\frac{\sum_{j=1}^{j=L} \left(R_j \left[(l_j + l_{j-1})^2 - l_{j-1}^2 \right] \right)}{\left(\sum_{j=1}^{j=L} l_j \right)^2} \right), \tag{4.3}$$

where $l_0 = 0$ and ρ is the Wi-Fi throughput efficiency factor, which characterizes the overhead of protocol negotiations and packet headers.

The Wi-Fi channel is shared by associated vehicles under a contention-based mechanism. With N vehicles sharing the Wi-Fi channel in a grid area, each vehicle achieves a data rate of $\overline{R}_{UAV}/N$. When the associated vehicles have a throughput requirement of R_{req}, the number of vehicles that can be simultaneously served by a UAV should be no more than $N_{U,\max}$, where

$$N_{U,\max} = \lfloor \overline{R}_{UAV} / R_{req} \rfloor. \tag{4.4}$$

CBS-to-Vehicle (C2V) Communications In this chapter, channel inversion power control is adopted for the CBS, where the transmit power is allocated based on channel conditions to ensure that the average SNR for all the associated vehicles keeps the same. The C2V channel gain is considered as $h_{i,C} = \varrho_{i,C} d_{i,C}^{-\alpha}$, where $\varrho_{i,C}$ is the channel fading following an exponential distribution with unit mean, $d_{i,C}$ is the distance between the vehicle and the CBS, and α is the pathloss exponent. Let B_C be the total available cellular bandwidth, which is equally shared by vehicles using C2V communications. $P_{C,\max}$ is the maximum available transmission power of the CBS. Given that the C2V channel gain varies with vehicle mobility, it is costly to perform real-time power allocation based on every vehicle's instantaneous channel condition. Thus, the average C2V channel gain is utilized for power allocation for vehicles in the same grid. The average C2V channel gain is dependent on the average C2V distance, which can be calculated referring to *Lemma 1* in [11]. When no UAVs are available in the system, the overall cellular network throughput at time slot t is

$$R_C(t) = \frac{B_C}{\sum_{v \in \mathcal{V}} d_{v,t}} \sum_{v \in \mathcal{V}} d_{v,t} \log\left(1 + \frac{P_{v,C}(t)h_{v,C}}{\frac{B_C \sigma^2}{\sum_{v \in \mathcal{V}} d_{v,t}}}\right), \tag{4.5}$$

where $P_{v,C}(t)$ and $h_{v,C}$ are the transmission power and average channel gain from the CBS to vehicles in grid v, and σ^2 is the noise power density. Thus, we have

$$\sum_{v \in \mathcal{V}} d_{v,t} P_{v,C}(t) = P_{C,\max}, \tag{4.6}$$

$$P_{v_1,C}(t)h_{v_1,C} = P_{v_2,C}(t)h_{v_2,C}, \ \forall v_1, v_2 \in \mathcal{V}.$$

Assuming that K UAVs have caching status **A**, delivery decision **S**, and trajectories **X**, the set of positions of the K UAVs at time slot t is $\mathcal{V}_t = \{x_{1,t}, x_{2,t}, \cdots, x_{K,t}\}$. In grid $x_{k,t}$, the average number of vehicles that need to be served by the CBS is

$$n_{C,k,t}^{A,X,S} = (1 - s_{k,t})d_{x_{k,t},t} + s_{k,t}d_{x_{k,t},t}\left(1 - \sum_{f_f \in \mathbf{a}_k} r_{x_{k,t},t,f}\right). \tag{4.7}$$

Denote by $B_{C,t}^{A,X,S}$ the cellular bandwidth allocated to each associated vehicle:

$$B_{C,t}^{A,X,S} = \frac{B_C}{\sum_{k=1}^{K} n_{C,k,t}^{A,X,S} + \sum_{u \in \mathcal{V}, u \notin \mathcal{V}_t} d_{u,t}}. \tag{4.8}$$

The average throughput achieved by the CBS (denoted by $R_{C,t}^{A,X,S}$) and UAV k (denoted by $R_{U,k,t}^{A,X,S}$) can be, respectively, expressed as

$$R_{C,t}^{A,X,S} = B_{C,t}^{A,X,S}\left(\sum_{u \in \mathcal{V}, u \notin \mathcal{V}_t} d_{u,t} \log\left(1 + \frac{P_{u,C}(t)h_{u,C}}{B_{C,t}^{A,X,S}\sigma^2}\right)\right.$$

$$\left. + \sum_{k=1}^{K} n_{C,k,t}^{A,X,S} \log\left(1 + \frac{P_{x_{k,t},C}(t)h_{x_{k,t},C}}{B_{C,t}^{A,X,S}\sigma^2}\right)\right),$$

$$R_{U,k,t}^{A,X,S} = \overline{R}_{UAV} \cdot \varepsilon \left(d_{x_k,t,t} \cdot s_{k,t} \cdot \sum_{f_f \in a_k} r_{x_k,t,t,f} \right), \tag{4.9}$$

where $\varepsilon(x)$ is a unit step function, $\varepsilon(x) = 1$ if $x > 0$; otherwise, $\varepsilon(x) = 0$. Then we can derive the overall system throughput as:

$$R(\mathbf{A}, \mathbf{X}, \mathbf{S}) = \sum_{t=1}^{T_U} \left(R_{C,t}^{A,X,S} + \sum_{k=1}^{K} R_{U,k,t}^{A,X,S} \right). \tag{4.10}$$

UAV Energy Consumption Models The energy consumption of the caching-enabled UAVs mainly includes two parts: the propulsion energy required for UAV movement and the communication energy for content delivery. Notice that the computing energy consumption in UAVs is not considered here since the computation-intensive tasks are carried out in the UAV control center instead of UAVs, which will be discussed with more details in Sect. 4.2.

Propulsion Energy For a rotary-wing UAV with speed V, the propulsion power and energy consumption during one time slot are, respectively, expressed as [11, 12]

$$P(V) = P_0 \left(1 + \frac{3V^3}{U^2} \right) + P_1 \left(\left(1 + \frac{V^4}{4v_r^4} \right)^{\frac{1}{2}} - \frac{V^2}{2v_r^2} \right)^{\frac{1}{2}} + \frac{1}{2} A V^3, \tag{4.11}$$

$$E_p(x) = \frac{x}{V} P(V) + \max \left\{ \Delta_t - \frac{x}{V}, 0 \right\} \cdot (P_0 + P_1),$$

where $x \in \{0, w, \sqrt{2}w\}$ is the flying distance within one time slot determined by the UAV trajectory planning, and P_0, P_1, U, v_r, and A are constant parameters related to the UAV's weight, wing area, air density, etc.

Communication Energy The probability that there are n vehicles requesting file f_f (with request probability $r_{v,t,f}$) in grid v at time slot t is

$$\Pr(f_f, d_{v,t}, n) = \binom{d_{v,t}}{n} r_{v,t,f}^n (1 - r_{v,t,f})^{d_{v,t}-n}. \tag{4.12}$$

Let a_k be the caching status of UAV k and $\mathcal{F}_k = \{f_{k,1}, f_{k,2}, \cdots, f_{k,m}\}$ be the set of m content files satisfying $a_{k,f} = 1$. Assuming that requests for different files are independent, the probability that there are n requests for files in $\mathcal{F}_k$ is

$$\Pr(\mathcal{F}_k, d_{v,t}, n) = \sum_{n_1=0}^{n} \Pr(f_{k,1}, d_{v,t}, n_1) \sum_{n_2=0}^{n-n_1} \Pr(f_{k,2}, d_{v,t}, n_2) \cdots$$

$$\sum_{n_{m-1}=0}^{n-\sum_{j=1}^{m-2} n_j} \Pr(f_{k,m-1}, d_{v,t}, n_{m-1})$$

$$\times \Pr(f_{k,m}, d_{v,t}, n - \sum_{j=1}^{m-1} n_j). \tag{4.13}$$

Note that one extreme case is that every vehicle requests all the cached files, in which case there are $m \cdot d_{v,t}$ requests for files in $\mathcal{F}_k$. For $n > m \cdot d_{v,t}$, we have $\Pr(\mathcal{F}_k, d_{v,t}, n) = 0$. Thus, given UAV k's caching status $\mathbf{a}_k$ and position $x_{k,t}$ at time t, the average communication energy consumption to serve the vehicle requests is

$$E_c(\mathbf{a}_k, x_{k,t}) = \sum_{n=1}^{m \cdot d_{v,t}} \Pr\left(\mathcal{F}_k, d_{x_{k,t},t}, n\right) P_U \min\left\{\Delta_t, \frac{n \cdot \varsigma_f}{R_{\text{UAV}}}\right\}. \qquad (4.14)$$

4.1.3 Problem Formulation

To maximize the overall network throughput in the mobile edge caching-assisted AGVNs under UAV energy constraints, $\mathbf{A}$, $\mathbf{S}$, and $\mathbf{X}$ should be jointly optimized since they are tightly intertwined. Based on the network throughput and UAV energy consumption analyzed in Sect. 4.1.2, the JCTO problem can be formulated as:

$$\text{(JCTO)} : \max_{\mathbf{A}, \mathbf{X}, \mathbf{S}} \quad R(\mathbf{A}, \mathbf{X}, \mathbf{S}) \qquad (4.15)$$

$$s.t. \quad \sum_{f_f \in \mathcal{F}} a_{k,f} \leq C_k, \quad \forall f_f \in \mathcal{F}, \forall k \in \mathcal{K}, \qquad (4.15a)$$

$$\sum_{t=1}^{T_U} E_p(\|x_{k,t} - x_{k,t-1}\|) + E_c(\mathbf{a}_k, x_{k,t}) \cdot s_{k,t} \leq E_{k,\max}, \qquad (4.15b)$$

$$x_{k,1} = x_{k,T_U} = v_{0,k}, \quad (x_{k,t-1}, x_{k,t}) \in \mathcal{E}, \qquad (4.15c)$$

$$a_{k,f} = \{0, 1\}, \quad s_{k,t} = \{0, 1\}, \qquad (4.15d)$$

where $E_{k,\max}$ is the on-board energy constraint for the k-th UAV. Constraints (4.15a) and (4.15b) restrict the maximum allowable number of files cached in each UAV and the maximum UAV energy consumption. Constraint (4.15c) means that UAVs should move along the edges in graph $\mathcal{G}$ and finally go back to the starting points. The JCTO problem in (4.15) is non-convex and intractable since $\mathbf{A}$, $\mathbf{X}$, and $\mathbf{S}$ should be jointly optimized for all the K UAVs under spatial–temporal network variations and UAVs' energy constraints.

4.2 Design of *LB-JCTO*

To solve the JCTO problem in (4.15), we propose a learning-based framework named *LB-JCTO*. As shown in Fig. 4.2, the *LB-JCTO* scheme includes two major stages: (1) offline optimization and (2) offline model training and online decision.

Offline Optimization In this stage, network information (including vehicle density and content request distribution) is obtained either from collected historical data or from predictions. Then we propose a CBTL algorithm to effectively solve the JCTO problem and achieve near-optimal joint solutions.

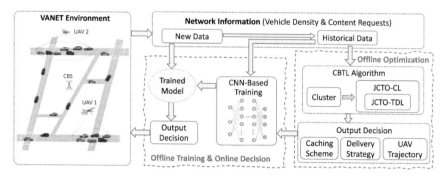

Fig. 4.2 Working diagram of the proposed *LB-JCTO* scheme

Offline Model Training and Online Decision   To enable real-time decision-making, AI-based methods can be utilized [13]. Specifically, we leverage a CNN-based DSL scheme to make fast decisions by learning from the CBTL algorithm. The CBTL algorithm works as a labeler (or supervisor) for the CNN-based learning model. Specifically, the network information and the solution obtained by the CBTL algorithm work together as labeled data, based on which a function mapping the input network information to output decisions can be learnt. After well trained, the learning model can be utilized to output the corresponding JCTO decisions rapidly to react to new network information. Meanwhile, the new network information can be collected and used to further train and update the learning model.

Notice that in the *LB-JCTO scheme*, network information collection, CBTL-based offline optimization, and CNN-based offline model training are carried out at the UAV control center or edge server with powerful computing and processing capabilities. Then the well-trained learning model can be transferred and implemented on UAVs to perform model inference locally to keep pace with the dynamic network information.

4.2.1   Offline Optimization

In the CBTL-based offline optimization stage, vehicles are grouped into $K + 1$ clusters, each of which is served by a UAV or the CBS. The clustering process ensures that each UAV flies within a constrained area to improve the energy efficiency and prevent potential collisions among UAVs. In vehicle clustering, a new metric combining three different types of similarities is considered in this chapter. Specifically, cellular throughput similarity, physical location similarity, and content preference similarity are considered to ensure that vehicles in the same cluster have similar C2V channel conditions, physical locations, and content interests.

After the vehicle clustering, the JCTO problem needs to be solved for the UAV in each cluster. Since the JCTO problem is non-convex and difficult to solve, we propose a CBTL algorithm by adopting a vertical decomposition that leads to the following two-layered structure of the problem:

- *Caching Layer (CL) Optimization:* The CL optimization problem can be reformulated as:

$$(\text{JCTO-CL}) : \max_{\mathbf{A}} \quad \sum_{t=1}^{T} R(\mathbf{A}, \mathbf{X}, \mathbf{S}) \tag{4.16}$$

$$s.t. \quad \text{Constraint (4.15a)}. \tag{4.16a}$$

- *Trajectory-and-Delivery Layer (TDL) Optimization:* The TDL optimization problem can be reformulated as:

$$(\text{JCTO-TDL}) : \max_{\mathbf{X}, \mathbf{S}} \quad \sum_{t=1}^{T} R(\mathbf{A}, \mathbf{X}, \mathbf{S}) \tag{4.17}$$

$$s.t. \quad \text{Constraints (4.15b)–(4.15d)}. \tag{4.17a}$$

In the proposed CBTL algorithm, the JCTO-CL subproblem is solved by using the PSO algorithm, and a time-based graph decomposition method is designed to solve the JCTO-TDL subproblem, which will be elaborated in Sects. 4.3.3 and 4.3.4. It is worth noting that when solving the JCTO-CL subproblem, the quality of a caching policy (i.e., the achievable $R(\mathbf{A}, \mathbf{X}, \mathbf{S})$) is determined by the optimal achievable performance with the JCTO-TDL subproblem. In other words, the JCTO-TDL optimization is embedded within the JCTO-CL optimization process in our CBTL-based offline optimization algorithm.

4.2.2 Offline Model Training and Online Decision

The CBTL-based offline optimization algorithm can achieve satisfactory throughput performance based on the given network information. However, in practice, it might be difficult to precisely predict the network information in each grid. For instance, a vehicle accident may significantly change the vehicle density in the collision location as well as in the surrounding grids. When the network condition changes unexpectedly after UAV dispatch, the achievable system throughput may decrease if the UAVs keep moving along the pre-designed trajectories. Under such circumstances, recalculating a good trajectory is critical to ensure efficient UAV service provision. Since UAVs are generally energy-constrained and have limited computing power, the proposed CBTL algorithm is not suitable to be implemented in UAVs to continuously update the trajectory and delivery decisions in real time. Therefore, in this chapter, we design a CNN-based DSL model that is trained offline under the supervision of the CBTL-based algorithm. Then UAVs can obtain the trained model from the edge server and perform real-time model inference locally [14].

CNN is effective in image processing and has been widely applied in many fields, e.g., image classification [15], text classification [16], traffic prediction [17], etc. With superior learning ability in image understanding, CNN is suitable for our problem for the following reasons. First, the convolutional layers can effectively capture the local dependencies and extract important features, e.g., network features like the road layout or dense areas with frequent requests. Second, CNN also introduces pooling mechanisms to reduce data dimension while preserving dominant features. With these two characteristics, the CNN-based model is not only good at learning features but also scalable to large-scale problems.

In the offline training of the CNN-based model, supervised learning is conducted to learn the matching function from the input data to the output decisions. Specifically, the input network information is labeled as channeled images and normalized before fed into the learning model. The optimized solutions obtained by the CBTL algorithm are labeled as model outputs to provide supervision to the learning model. The detailed learning model structure will be introduced in Sect. 4.4.

4.3 CBTL-Based Offline Optimization

4.3.1 Determining the Number of UAVs

Basically, with UAV-assisted communications in the AGVN, the network throughput can be improved and thus each vehicle's service quality is enhanced. From the users' point of view, more UAVs should be dispatched into the system to improve the service satisfaction levels. The service providers, on the other hand, prefer to use the least resource to achieve the best gain and focus more on cost efficiency, e.g., the performance enhancement introduced by each UAV. Figure 4.3 shows the impact of the number of UAVs on users' satisfaction level (approximated by using sigmoid function [18]) and throughput improvement introduced by each UAV. We can see that the average user satisfaction level increases, while the throughput improvement efficiency decreases with an increasing number of UAVs. Thus, the optimal K varies in different scenarios with diverse user satisfaction requirements, the number of UAVs owned by the provider, and/or the deployment cost efficiency constraint.

In our caching-assisted AGVNs, we aim to minimize the number of UAVs required to satisfy the vehicles' throughput requirement R_{req}. When adopting the channel inversion power control, all vehicles served by the CBS have the same received signal strength γ_C, which can be calculated based on (4.6)

$$\gamma_C = P_{C,\max} \bigg/ \left(\sum_{v \in \mathcal{V}} \frac{d_{v,t}}{h_{v,C}} \right). \tag{4.18}$$

When satisfying the vehicles' throughput requirement, the CBS can serve more vehicles if it serves close vehicles rather than remote vehicles, since the close ones have better C2V links and require smaller transmit power. Therefore, we sort the

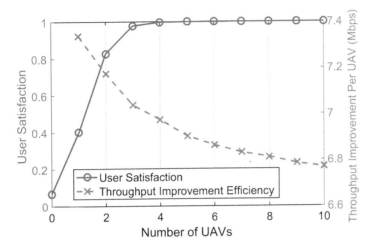

Fig. 4.3 Impact of K on user satisfaction level and throughput improvement efficiency. $H = 35$ m, $P_{C,\text{max}} = 43$ dBm, $B_C = 20$ MHz, and $P_U = 28$ dBm

grids in descending order based on C2V channel power gain. Let v_i be the i-th closest grid to the CBS. When the CBS serves users in the first N_C nearest grids, the achievable throughput per vehicle should satisfy

$$\frac{B_C}{\sum_{i=1}^{N_C} d_{v_i,t}} \log \left(1 + \frac{P_{C,\text{max}} \sum_{i=1}^{N_C} d_{v_i,t}}{B_C \sigma^2 \sum_{i=1}^{N_C} \frac{d_{v_i,t}}{h_{v_i,C}}} \right) \geq R_{req}. \tag{4.19}$$

The left side of Eq. (4.19) decreases monotonously with N_C. Let $N_{C,\text{max}}$ denote the maximum value of N_C satisfying Eq. (4.19). Although the closed-form expression of the optimal N_C cannot be derived, $N_{C,\text{max}}$ can be easily calculated by using approaches like bisection method. Thus, the minimum number of UAVs required to ensure vehicle throughput requirement is expressed as

$$K_{\min} = \left\lceil \frac{\sum_{v \in \mathcal{V}} d_{v,t} - \sum_{i=1}^{N_{C,\text{max}}} d_{v_i,t}}{N_{U,\text{max}}} \right\rceil. \tag{4.20}$$

4.3.2 Vehicle Clustering

With K UAVs dispatched into the AGVN, vehicles can be clustered into $K + 1$ groups to be served by the UAVs and the CBS. In this chapter, a widely used clustering method, K-means clustering [19], is adopted by considering the following similarities:

1. *Cellular Throughput Similarity:* When a UAV is dispatched into the AGVN, the throughput gain increases if it serves vehicles with poor cellular performance [20]. Therefore, the cellular throughput similarity should be considered to cluster vehicles to be served by UAVs. In other words, vehicles with good C2V channels prefer to be in the same cluster and served by the CBS, while other vehicles need to be served by UAVs. When assigned the same cellular bandwidth B_0 and transmit power P_0, vehicles in grid v achieve a throughput of $R_{C,v} = B_0 \log(1 + P_0 h_{v,C}/\sigma^2)$. The cellular throughput similarity between vehicles in grids v and u is evaluated by

$$\text{sim}_{u,v,CT} = \min\left\{\frac{R_{C,v}}{R_{C,u}}, \frac{R_{C,u}}{R_{C,v}}\right\} \in [0, 1]. \tag{4.21}$$

2. *Physical Location Similarity:* Basically, vehicles served by the same UAV should be in proximity to avoid extra UAV propulsion energy consumption. Thus, the physical location similarity among vehicles in grids v and u is evaluated by

$$\text{sim}_{u,v,PL} = 1 - \frac{\text{dist}_{u,v}}{\max_{u,v \in \mathcal{V}} \text{dist}_{u,v}} \in [0, 1], \tag{4.22}$$

where $\text{dist}_{u,v}$ is the average distance between grids u and v.

3. *Content Preference Similarity:* To improve the UAV caching resource utilization efficiency, vehicles with similar content interests should be grouped and served by the same UAV. Utilizing cosine similarity to evaluate the file preference similarity [9], we express the content preference similarity between grids u and v during T_U time slots as

$$\text{sim}_{u,v,CP} = \frac{1}{T_U} \sum_{t=1}^{T_U} \frac{\mathbf{r}_{v,t} \cdot \mathbf{r}_{u,t}}{\|\mathbf{r}_{v,t}\| \cdot \|\mathbf{r}_{u,t}\|}, \tag{4.23}$$

where $\mathbf{r}_{v,t} = [r_{v,t,f_1}, \cdots, r_{v,t,f_F}]$ is the request distribution in grid v at time t and $\text{sim}_{u,v,CP} \in [0, 1]$.

Taking the above-mentioned three metrics into account, the overall similarity between grids u and v can be evaluated by

$$\text{sim}_{u,v}^{all} = \text{sim}_{u,v,CT}^{\alpha_1} \cdot \text{sim}_{u,v,PL}^{\alpha_2} \cdot \text{sim}_{u,v,CP}^{\alpha_3}, \tag{4.24}$$

where parameters α_1, α_2, and α_3 control the relative importance of the three metrics. For example, a large value of α_1 means that the cellular performance similarity has a high priority, while a small α indicates that we put more emphasis on the other two similarity metrics. Targeting at maximizing $\text{sim}_{u,v}^{all}$ within a cluster, the K-means-based clustering algorithm can be summarized as shown in **Algorithm 3.**

Algorithm 3: K-means-based vehicle clustering

Let $\overline{d}_v = \frac{1}{T_U} \sum_{t=1}^{T_U} d_{v,t}, \overline{\mathbf{r}}_v = \frac{1}{T_U} \sum_{t=1}^{T_U} \mathbf{r}_{v,t}$.

Step 1: Centroid Initialization:
The first cluster centroid v_C^0 is the grid where the CBS is located.
Randomly choose K grids, $v_C^1, \cdots, v_C^K$, as the centroids for the remaining K clusters.
Step 2: Grid Clustering: $K + 1$ clusters, denoted by $C_0, C_1, \cdots, C_K$, are created by associating every grid with the centroid with the maximum similarity based on (4.24).
Step 3: Centroid Update: Update the $K + 1$ cluster centroids:

$$v_C^0 = v_C^0, \quad v_C^k = \frac{1}{\sum_{v \in C_k} \overline{d}_v} \sum_{v \in C_k} \overline{d}_v v,$$

$$R_{C,v_C^k} = \frac{1}{\sum_{v \in C_k} \overline{d}_v} \sum_{v \in C_k} \overline{d}_v R_{C,v}, \quad \mathbf{r}_{v_C^k} = \frac{1}{\sum_{v \in C_k} \overline{d}_v} \sum_{v \in C_k} \overline{d}_v \overline{\mathbf{r}}_v.$$

Step 4: Repeat **Steps 2–3** until converging.
Step 5: Repeat **Steps 1–4** and choose the best for multiple runs.

4.3.3 JCTO-TDL Optimization in the CBTL Algorithm

After vehicle clustering, the JCTO problem is solved for each UAV based on the problem decomposition as mentioned in Sect. 4.2.1. In this part, the JCTO-TDL subproblem is addressed by the proposed time-based graph decomposition method.

Given the caching policy and the current position of a UAV, when the UAV flies to any feasible position in the next time slot, the achievable network throughput and the corresponding energy consumption can be calculated according to Sect. 4.1.2. The JCTO-TDL subproblem, which aims to find the optimal content delivery and UAV trajectory to maximize the network throughput under the energy constraint, is similar to the RCSP problem, which has been investigated in [5]. However, the RCSP algorithm cannot be directly applied to the JCTO-TDL subproblem for the following reasons: (1) at each time slot, $s_{k,t}$ can be either 0 or 1, which corresponds to two edges connecting the adjacent grids with different weights (i.e., achievable network throughput) and costs (i.e., energy consumption), as shown in Fig. 4.4a, and (2) due to the time-variant $\mathbf{D}$ and $\mathbf{R}$, the weight and cost of an edge change when visited by UAVs at different time slots. As shown in Fig. 4.4b, with two trajectories from v_1 to v_4 (respectively, marked in red and blue), the throughput and energy consumption vary when the UAV flies from v_3 to v_4 at different time slots.

To address the above-mentioned issues, we propose a time-based decomposition method to expand graph $\mathcal{G}$ into a directed graph, as shown in Fig. 4.4c. The edge connecting v_i at time t_1 and v_j at time t_2 exists only if $t_2 = t_1 + \Delta_t$ and $(v_i, v_j) \in \mathcal{V}$. An example path is depicted as the purple curve in Fig. 4.4c, representing a possible UAV trajectory and content delivery decision in each step from the source (v_1 at

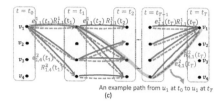

Fig. 4.4 A simple example of JCTO-TDL design with time-based graph decomposition. ($R_{i,j}^1(t)$ and $R_{i,j}^2(t)$ denote the achievable throughput when flying from v_i to v_j at time t with and without content delivery, and $e_{i,j}^1(t)$ and $e_{i,j}^2(t)$ are the corresponding energy consumption). (a) Two paths between every two nodes showing different content delivery cases. (b) Graph with edges that have time-variant weights and costs. (c) Trajectory and content delivery design with time-based decomposed graph

---

Algorithm 4: JCTO-TDL optimization in CBTL algorithm

---

v_k: the starting and ending points of UAV k.
Step 1: For any edge $(u, v) \in \mathcal{E}$ and $t \in [1, T_U]$, calculate the edge weight (throughput) and cost (energy consumption).
Step 2: Use shortest path (SP) algorithms (e.g., Dijkstra's algorithm) to find the path with smallest cost.
Let E_{t_0, t_T} denote the sum cost from source to v_k at time slot t_T.
Find t_T such that $E_{t_0, t_T} \leq E_{k,all}$ and $E_{t_0, t_{T+1}} > E_{k,all}$. Let $t_T^{max} = t_T$.
Step 3: Use SP algorithms to find the path with largest cost.
Find t_T such that $E_{t_0, t_T} \leq E_{k,all}$ and $E_{t_0, t_{T+1}} > E_{k,all}$. Let $t_T^{min} = t_T$.
for $t_T = [t_T^{min}, t_T^{max}]$ **do**
    Construct time-based decomposed graph as shown in Fig. 4.4c.
    Apply RCSP algorithms to find the optimal path with maximized sum weight.
    Record the best path which leads to the maximum achievable network throughput.
end
Output the recorded best path.

---

time t_0) to the destination (v_1 at time t_T.[4]) To this end, the JCTO-TDL subproblem is equivalent to finding the optimal path in the expanded graph to maximize sum weights under the cost constraint. With given source and destination, RCSP algorithms can be leveraged to find the optimal path. However, considering that the UAV's energy consumption varies with different trajectories and delivery decisions, it is difficult to determine the destination t_k when the UAV exhausts its energy. To address this issue, we execute an energy-constrained line search on t_k to find the optimal JCTO-TDL solution, as described in **Algorithm 4**.

---

[4] The UAV returns to the UAV control center at time $t_T \leq T_U$. From time slot t_{T+1} to T_U, the UAV stays in the UAV center without content delivery and charges its battery for the next flight.

4.3.4 JCTO-CL Optimization in the CBTL Algorithm

After the JCTO-TDL optimization, the optimal $\mathbf{X}$, $\mathbf{S}$, and the corresponding $R(\mathbf{A}, \mathbf{X}, \mathbf{S})$ can be obtained with given caching policy. In this subsection, the content placement is optimized to further improve the network throughput. However, conventional linear programming approaches cannot solve the JCTO-CL subproblem since the closed-form expression for $R(\mathbf{A}, \mathbf{X}, \mathbf{S})$ cannot be derived. Compared to the exhaustive searching scheme with exorbitant time complexity, heuristic algorithms, especially the evolutionary heuristics, are considered as better alternative choices to obtain the optima [21]. Specifically, we utilize the PSO algorithm to solve the JCTO-CL subproblem due to its low computational cost and fast convergence [22].

When applying the PSO algorithm, a group of particles are generated, each of which has a position representing a potential caching policy. Then the fitness values (achievable network throughput) of these particles are calculated based on the analysis in Sect. 4.3.3. Based on the particles' positions and fitness values, there exist a local optimal position ($\varpi_{\mathrm{local}}^{\ell}(t)$) for each particle ℓ and a global optimal position ($\varpi_{\mathrm{global}}(t)$) for the entire particle swarm at the t-th iteration. Then the position $\varpi^{\ell}(t+1)$ and velocity $v^{\ell}(t+1)$ of particle ℓ are updated as

$$v^{\ell}(t+1) = \phi v^{\ell}(t) + c_1\phi_1(\varpi_{\mathrm{local}}^{\ell}(t) - \varpi^{\ell}(t)) + c_2\phi_2(\varpi_{\mathrm{global}}(t) - \varpi^{\ell}(t)), \tag{4.25}$$

$$\varpi^{\ell}(t+1) = \varpi^{\ell}(t) + v^{\ell}(t+1),$$

where ϕ determines convergence speed, c_1 and c_2 are local and global learning coefficients, and ϕ_1 and ϕ_2 are positive random variables. The iteration terminates when a termination criterion (e.g., reaching the maximum iterations or minimum error criteria) is met.

To this end, with given ground vehicle densities and content request distributions, the JCTO problem can be solved effectively by our proposed CBTL algorithm.

4.4 CNN-Based Learning for Online Decision

Based on the optimization results obtained by the CBTL algorithm, a CNN-based DSL scheme is designed in this section to make real-time decisions under dynamic network conditions.

4.4.1 Image-Like Input Data

As discussed in Sect. 4.3, the JCTO problem is investigated with dynamic network information (e.g., $\mathbf{D}$ and $\mathbf{R}$). In this part, we adopt an image-based method to present the spatial–temporal network dynamics as images to facilitate the learning scheme.

Vehicle density $\mathbf{D}$ is a three-dimensional array consisting of T_U two-dimensional matrices, each of which can be viewed as a channel of an image with $N_{\text{row}} \times N_{\text{col}}$ elements. In this way, each element in the matrix can correspond to a pixel in the image. The input vehicle density can then be considered as an image with T_U channels, which differs from traditional images which commonly have three channels, i.e., RGB.

Content request distribution $\mathbf{R}$, on the other hand, is a four-dimensional array. To keep the input dimension consistent without losing useful information on the request distribution, we use $\phi_{v,t}$ (as discussed in Sect. 4.1.1, $v = (i, j), i \in [1, N_{\text{row}}], j \in [1, N_{\text{col}}]$) to characterize the content request distribution in grid v at time t. Then the three-dimensional array $\mathbf{\Phi}_{N_{\text{row}} \times N_{\text{col}} \times T_U}$, with the (i, j, t)-th entry being $\phi_{i,j,t}$, can also be treated as an image with T_U channels.

It is worth noting that $\mathbf{D}$ and $\mathbf{\Phi}$ have different scales. Considering that neural networks are sensitive to the scaling and distribution of their inputs, proper normalization is critical for convergence [23]. In our CNN-based learning model, the input data is normalized before fed into the learning model by using the min–max normalization as follows:

$$x' = \frac{x - \min(x)}{\max(x) - \min(x)}. \tag{4.26}$$

With normalized range of the input data, the impact of $\mathbf{D}$ and $\mathbf{\Phi}$ is rescaled to approximately the same level to facilitate the learning process.

4.4.2 CNN-Based Model Training

Figure 4.5 shows the structure of the CNN-based learning model, which includes three main parts, i.e., model input, feature extraction, and decision-making and output.

1. Model input contains the image-like network information arrays with spatial–temporal characteristics, as explained in Sect. 4.4.1. Thus, the l-th input data can be written as

$$\text{IN}_l = \left[\mathbf{d}_1, \cdots, \mathbf{d}_t, \cdots, \mathbf{d}_{T_U}, \boldsymbol{\phi}_1, \cdots, \boldsymbol{\phi}_t, \cdots, \boldsymbol{\phi}_{T_U} \right], \tag{4.27}$$

where $\mathbf{d}_t$ and $\boldsymbol{\phi}_t$ are $N_{\text{row}} \times N_{\text{col}}$ matrices with (i, j)-th entry being $d_{i,j,t}$ and $\phi_{i,j,t}$, respectively.

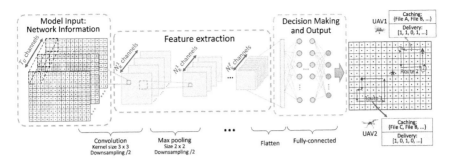

Fig. 4.5 Structure of the CNN-based DSL model

2. The network features are extracted by utilizing the convolutional and pooling layers, which are the core part of the CNN model. In the ℓ-th convolutional layer, there are N_k^ℓ kernels (or filters) of size $W_k^\ell \times H_k^\ell$ with stride s_k^ℓ. For example, as shown in the first convolutional layer in Fig. 4.5, there are N_k^1 kernels of size 3×3 with stride 2, and then the input network information of size $2N_{\text{row}} \times N_{\text{col}} \times T_U$ is fed into the first layer to convolve with the N_k^1 kernels, producing an output of size $\left(\lfloor \frac{2N_{\text{row}}-3}{2} \rfloor + 1 \right) \times \left(\lfloor \frac{N_{\text{col}}-3}{2} \rfloor + 1 \right) \times N_k^1$. The output is then activated by an activation function (such as ReLU, Sigmoid, and Softmax), which introduces nonlinearity into the system to improve the model's learning capability. After the convolutional layer, a pooling layer is introduced to downsample the convolved features, by using either a max-pooling function or an average-pooling function. For example, in the second layer (first pooling layer) in Fig. 4.5, we use max-pooling with size 2×2 and stride 2 to downsample the convolved features and the output is of size $\left(\lfloor \frac{\lfloor \frac{2N_{\text{row}}-3}{2} \rfloor - 1}{2} \rfloor + 1 \right) \times \left(\lfloor \frac{\lfloor \frac{N_{\text{col}}-3}{2} \rfloor - 1}{2} \rfloor + 1 \right) \times N_k^1$. With max-pooling layers, the data dimension can be reduced whereas dominant features can be preserved, and the level of distortion invariance can be improved. Generally, the CNN has multiple convolutional and pooling layers to capture not only some straightforward features (e.g., detection of road layout) but also sophisticated features (e.g., recognizing needy areas with intensive requests or undesired C2V links) of the model input.

3. In the decision-making part, the extracted features are first flattened and concatenated into a column vector. Then some fully connected layers are added to learn the combinations of the extracted network features and out decisions. Activation functions are added after each fully connected layer to introduce nonlinearity for improving learning capability.

   Over a series of training epochs, a possibly nonlinear function mapping the input and output can be learnt with the CNN-based model. The output of the CNN-based model is represented by a column with length $\sum_k C_k + K \cdot (T_U + T_U \cdot 2)$, which includes the following:

- $\sum_k C_k$ numbers in the output represent the UAVs' caching status, with each number in range $[1, F]$ showing the index of the files being cached.
- $K \cdot T_U$ numbers in the output represent the K UAVs' delivery decisions, with each number in $\{0, 1\}$.
- $K \cdot T_U \cdot 2$ numbers in the output describe the trajectories of the K UAVs in T_U time slots, with each tuple of two numbers $(i, j) \in [1, N_{row}] \times [1, N_{col}]$ showing the location of a UAV.

4.5 Performance Evaluation

4.5.1 Experimental Settings

We conduct extensive trace-driven simulations to evaluate the proposed *LB-JCTO* scheme. We adopt the Didi Chuxing GAIA Initiative dataset, which includes taxi GPS traces within the second ring road in Xi'an [24]. The dataset logs key attributes of vehicular mobility including vehicle positions, vehicle ID, and corresponding timestamps. The traffic data is aggregated every 2–4 s from 1 October 2016 to 31 October 2016 (31 days). Specifically, we focus on a 2×2 km square area within longitude range $(108.9169, 108.9300)$ and latitude range $(34.2290, 34.2466)$. For UAV communications, the zone parameters and the corresponding data rates are calculated referring to [11]. The default values of main parameters related to the UAVs are $K = 2$, $H = 30$ m, $V = 15$ m/s, $E_{k,max} = 50$ KJ, $\xi_{LoS} = 0.99$, and $\gamma_{max} = 37.5$ dB. For cellular communications, the default parameters are $\alpha = 3$, $B_C = 20$ MHz, $P_{C,max} = 20$ W, and $\sigma^2 = 10^{-15}$ W/Hz. The Zipf exponent ξ is set to be 0.7 and the time slot length Δ_t is set to be 5 s unless otherwise specified. We implement a learning model with 11 layers detailed as follows. The model has three convolutional layers with channel sizes of 16, 32, and 64, respectively. The kernel sizes and strides for each convolutional layer are $(3, 3)$ and $(2, 2)$, respectively. After each convolutional layer, a max-pooling layer is added with pool size $(2, 2)$. Four fully connected layers are then added with 1024, 1024, 512, and 256 neurons, respectively, followed by one output layer. ReLU activation function is added after each layer to introduce nonlinearity and improve learning capability.

4.5.2 Evaluation of CBTL-Based Offline Optimization

To evaluate the effectiveness of the offline CBTL-based optimization algorithm, the following benchmark schemes are considered:

- *Exhaustive Search (ES) Algorithm:* An ES method is used in JCTO-CL to optimize the content placement decision.

- *Greedy Algorithm:* UAVs fly and deliver content greedily in JCTO-TDL optimization, where the UAVs always visit nearby locations with the best throughput performance in each step. The UAVs return to the starting points when the remaining energy is barely sufficient for returning.

Figure 4.6 shows the network throughput and execution time with the PSO- and ES-based algorithms to solve the JCTO-CL subproblem. As shown in Fig. 4.6a, applying the PSO-based method in our CBTL algorithm achieves almost the same throughput performance as applying the ES method. However, the PSO-based method is much less time-consuming than the ES algorithm, as shown in Fig. 4.6b. Therefore, instead of learning from the optimal ES algorithm, the CNN-based model utilizes the CBTL algorithm as the learning supervisor to save substantial time in data labeling such that more data can be used to train the model, which is more energy-efficient.

Figure 4.7 shows the network throughput and the corresponding execution time of applying RCSP-based and greedy-based algorithms with different values of Δ_t and H. A small Δ_t means a fine-grained CBTL optimization in time domain,

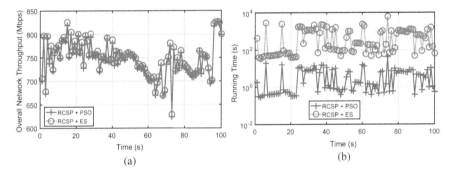

Fig. 4.6 Comparison between PSO- and ES-based algorithms. (**a**) Network throughput. (**b**) Simulation running time

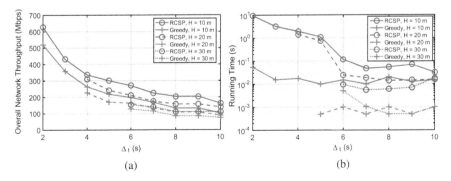

Fig. 4.7 Comparison between RCSP- and greedy-based algorithms. (**a**) Network throughput. (**b**) Simulation running time

whereas a large Δ_t (e.g., Δ_t equals the UAV endurance time and $T_U = 1$) is more related to the case with UAV deployment instead of trajectory design. Besides, with the simulation setting in Fig. 4.7, a lower UAV operation height leads to a smaller coverage area and indicates a refined division of the target area in spatial domain. We can also see that the network throughput and the execution time both increase with smaller Δ_t and H, which can be attributed to the more sophisticated design in our CBTL algorithm. Focusing on the network throughput, we can see from Fig. 4.7a that applying RCSP-based method in the CBTL algorithm achieves a better performance. However, its time complexity is higher than that of the greedy algorithm, especially in fine-grained optimization cases with small Δ_t and H, as shown in Fig. 4.7b. From Figs. 4.6 and 4.7, we can conclude that the proposed CBTL optimization algorithm can achieve near-optimal network throughput performance with time complexity slightly higher than the greedy-based algorithm.

Figure 4.8 shows the impact of the number of UAVs K and the UAV onboard energy capacity $E_{k,\max}$ on the achievable network throughput. To further illustrate the effectiveness of the proposed scheme, we also compare the case of UAV deployment. For the fixed UAV deployment case, the optimal UAV positions

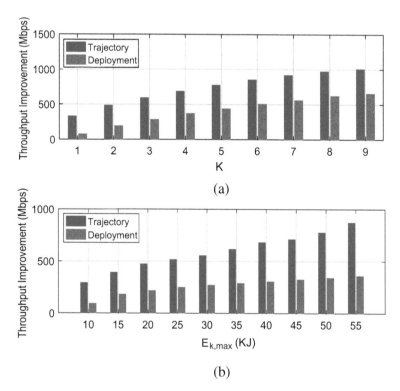

Fig. 4.8 Throughput performance vs. K and $E_{k,\max}$. (**a**) Impact of number of UAVs. (**b**) Impact of available UAV on-board energy

are determined based on the network conditions at the first slot and then UAVs hover in those positions until energy depletion. As shown in Fig. 4.8a, the overall network throughput improvement increases with more UAVs dispatched into the AGVN. However, throughput improvement introduced by each UAV diminishes with increasing K. The UAV trajectory design case outperforms the UAV deployment case, since the former is able to capture the network dynamics to ensure effective content delivery. In addition, as shown in Fig. 4.8b, a larger $E_{k,\max}$ leads to a higher network throughput for both UAV trajectory design and deployment cases since the UAVs can stay in the system for a longer time. We can also see from the figure that the throughput performance gap between the UAV trajectory and deployment cases increases with $E_{k,\max}$, since UAV trajectory design scheme is adaptive to the network variance and enables effective utilization of the energy to provide delivery services.

4.5.3 Evaluation of EI-Based CNN Learning Model

In this subsection, the achievable performance of the CNN-based learning model is evaluated. Specifically, we have trained multiple models by using different sets of trace data for performance comparison. In the simulation results, "Model: [20, 50]," "Model: [50, 80]," "Model: $\geq$ 80," and "Model: General" are used to represent the models that are trained by using data where the number of vehicles in the target scenario is between 20 and 50, between 50 and 80, no less than 80, and unconstrained, respectively. For all the experiments, we adopt the trace data that is never used in the model training process to test the performance.

In Fig. 4.9, network throughput and execution time of the CNN-based learning model are presented by using vehicle trace data during 9:15~10:30AM on 1 October 2016. The real-time number of vehicles in the target area is depicted in Fig. 4.9a. When using the general model to make online decisions, as shown in Fig. 4.9b, although the network throughput of the CNN model outperforms that of the greedy-based algorithm, it is not as satisfactory as the RCSP-based offline CBTL algorithm. Thus, training one model to incorporate all the features in different network conditions where the vehicle density varies from 40 to 120 may not be the best solution. For performance comparison, in Fig. 4.9c, the "Model: $\geq$ 80" is used to make JCTO decisions. Although the throughput performance is far from ideal in the beginning, the CNN-based learning model can achieve almost the same network throughput as the RCSP-based CBTL algorithm when the vehicle density increases over 80 (shown within the red rectangle). More importantly, as shown in Fig. 4.9d, the CNN-based learning model is much less time-consuming when compared with the CBTL-based algorithms. Therefore, a well-trained CNN-based model can make online decisions with a satisfactory throughput performance and low complexity. However, how to select and refine the models for different network conditions requires further investigation, to achieve the best learning performance.

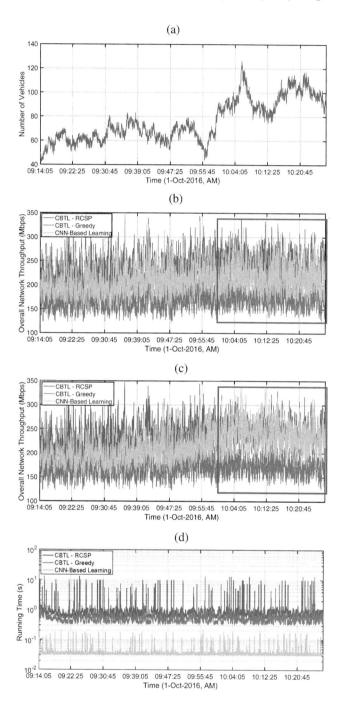

Fig. 4.9 Performance for CNN-based online decision model. (**a**) Vehicle density. (**b**) Network throughput with "Model: General". (**c**) Network throughput with "Model: ≥ 80". (**d**) Simulation running time with "Model: ≥ 80"

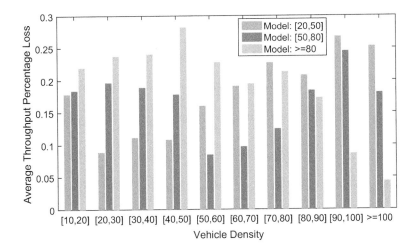

Fig. 4.10 Throughput performance with density-related CNN models

The achievable performance of different CNN-based models under different density conditions is shown in Fig. 4.10. Specifically, average throughput percentage loss is used as a metric to evaluate the performance.[5] As shown in Fig. 4.10, "Model: [20, 50]" achieves the best performance (i.e., with the lowest throughput percentage losses) when applied to scenarios with vehicle densities being [20, 50], but its performance is unsatisfactory in other cases. Similar results can be seen for "Model: [50, 80]" and "Model: $\geq$ 80." Thus, training multiple density-specified models and applying them in corresponding scenarios is a decent method to enhance model performance. However, a fine-grained model training can introduce significant training and storage cost. Besides, if the model granularity is too fine, the performance will be impacted since less training data is available. Therefore, the optimal number of learning models should be determined based on the throughput requirements, computing and storage capacities of devices, data availability, and so on.

The impact of training data on the performance of the CNN-based models is demonstrated in Fig. 4.11. The X-axis represents the cases where 0.1~0.9 of the available data is selected for model training. "Uniform" and "Continuous" represent that the training data is selected uniformly and continuously from the available dataset, respectively. In "Fixed Interval" case, the training data is selected at fixed intervals, e.g., choosing the first three out of every ten pieces of data. As shown in Fig. 4.11, with more data used for model training, a better throughput performance can be achieved because more information about the network features

---

[5] Throughput percentage loss is defined as $\frac{\hat{R}-\tilde{R}}{\hat{R}}$, where $\hat{R}$ and $\tilde{R}$ are the achievable network throughput of the CBTL-based optimization algorithm and the CNN-based learning model, respectively.

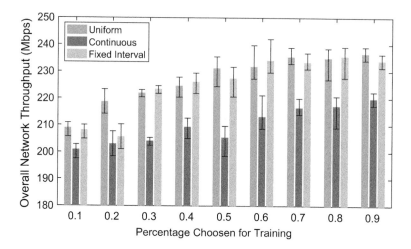

Fig. 4.11 Network throughput with different methods of training data selection

and regularities can be learnt with the CNN-based model. Among the three different methods of training data selection, the "Continuous" case achieves the worst performance since the training data is highly temporally correlated, which makes it inefficient to learn the network dynamics in different time intervals. On the other hand, the "Uniform" and "Fixed Interval" behave well since they are able to capture the network variance in the time domain. Thus, to obtain a well-performed learning model, one should gather and select enough training data to learn as many potential features as possible.

4.6 Summary

In this chapter, we have investigated the joint design of UAV caching and trajectory in the mobile edge caching-assisted AGVNs. As the formulated JCTO problem is non-convex and difficult to solve in a timely manner, we have proposed *LB-JCTO* to offline optimize the JCTO problem and train a learning model at the edge to make online decisions. Specifically, in the offline stage, the CBTL algorithm has been devised to solve the JCTO problem. Then a CNN-based DSL model is trained to learn the CBTL algorithm and make fast decisions in the online stage. Extensive trace-driven experiments have been conducted to demonstrate that the CBTL algorithm can efficiently solve the JCTO problem, and the CNN-based learning model can well emulate the capability of CBTL while satisfying the real-time requirements.

References

1. F. Lyu, H. Zhu, N. Cheng, H. Zhou, W. Xu, M. Li, X. Shen, Characterizing urban vehicle-to-vehicle communications for reliable safety applications. IEEE Trans. Intell. Transp. Syst. **21**(6), 2586–2602 (2020)
2. M. Noor-A-Rahim, Z. Liu, H. Lee, G.M.N. Ali, D. Pesch, P. Xiao, A survey on resource allocation in vehicular networks. IEEE Trans. Intell. Transp. Syst. (2020). https://doi.org/10.1109/TITS.2020.3019322
3. H. Wu, J. Chen, W. Xu, N. Cheng, W. Shi, L. Wang, X. Shen, Delay-minimized edge caching in heterogeneous vehicular networks: a matching-based approach. IEEE Trans. Wireless Commun. **19**(10), 6409–6424 (2020)
4. J. Chen, H. Wu, P. Yang, F. Lyu, X. Shen, Cooperative edge caching with location-based and popular contents for vehicular networks. IEEE Trans. Veh. Technol. **69**(9), 10291–10305 (2020)
5. G.Y. Handler, I. Zang, A dual algorithm for the constrained shortest path problem. Networks **10**(4), 293–309 (1980)
6. Y. LeCun, Y. Bengio, G. Hinton, Deep learning. Nature **521**(7553), 436–444 (2015)
7. F. Ono, H. Ochiai, R. Miura, A wireless relay network based on unmanned aircraft system with rate optimization. IEEE Trans. Wireless Commun. **15**(11), 7699–7708 (2016)
8. E. Baştuğ, M. Bennis, E. Zeydan, M.A. Kader, I.A. Karatepe, A.S. Er, M. Debbah, Big data meets telcos: a proactive caching perspective. J. Commun. Netw. **17**(6), 549–557 (2015)
9. B. Chen, C. Yang, Caching policy for cache-enabled D2D communications by learning user preference. IEEE Trans. Commun. **66**(12), 6586–6601 (2018)
10. W. Shi, J. Li, N. Cheng, F. Lyu, S. Zhang, H. Zhou, X. Shen, Multi-drone 3-D trajectory planning and scheduling in drone-assisted radio access networks. IEEE Trans. Veh. Technol. **68**(8), 8145–8158 (2019)
11. H. Wu, J. Chen, F. Lyu, L. Wang, X. Shen, Joint caching and trajectory design for cache-enabled UAV in vehicular networks, in *Proc. IEEE 11th International Conference on Wireless Communications and Signal Processing (WCSP), Xi'an* (2019), pp. 1–6
12. Y. Zeng, J. Xu, R. Zhang, Energy minimization for wireless communication with rotary-wing UAV. IEEE Trans. Wireless Commun. **18**(4), 2329–2345 (2019)
13. X. Shen, J. Gao, W. Wu, K. Lyu, M. Li, W. Zhuang, X. Li, J. Rao, AI-assisted network-slicing based next-generation wireless networks. IEEE Open J. Veh. Technol. **1**(1), 45–66 (2020)
14. Z. Zhou, X. Chen, E. Li, L. Zeng, K. Luo, J. Zhang, Edge intelligence: paving the last mile of artificial intelligence with edge computing. Proc. IEEE **107**(8), 1738–1762 (2019)
15. A. Krizhevsky, I. Sutskever, G.E. Hinton, Imagenet classification with deep convolutional neural networks, in *Proceedings of Advances in Neural Information Processing Systems, Lake Tahoe* (2012), pp. 1097–1105
16. S. Lai, L. Xu, K. Liu, J. Zhao, Recurrent convolutional neural networks for text classification, in *Proc. Twenty-Ninth AAAI Conference on Artificial Intelligence, Austin* (2015), pp. 2267–2273
17. X. Ma, Z. Dai, Z. He, J. Ma, Y. Wang, Y. Wang, Learning traffic as images: a deep convolutional neural network for large-scale transportation network speed prediction. Sensors **17**(4), 818 (2017)
18. E.B. Rodrigues, F.R.M. Lima, T.F. Maciel, F.R.P. Cavalcanti, Maximization of user satisfaction in OFDMA systems using utility-based resource allocation. Wireless Commun. Mob. Comput. **16**(4), 376–392 (2016)
19. T. Kanungo, D.M. Mount, N.S. Netanyahu, C.D. Piatko, R. Silverman, A.Y. Wu, An efficient k-means clustering algorithm: analysis and implementation. IEEE Trans. Pattern Anal. Mach. Intell. **24**(7), 881–892 (2002)
20. F. Lyu, P. Yang, H. Wu, C. Zhou, J. Ren, Y. Zhang, X. Shen, Service-oriented dynamic resource slicing and optimization for space-air-ground integrated vehicular networks. IEEE Trans. Intell. Transp. Syst. (2021). https://doi.org/10.1109/TITS.2021.3070542

21. L. Liu, Y. Song, H. Zhang, H. Ma, A.V. Vasilakos, Physarum optimization: a biology-inspired algorithm for the Steiner tree problem in networks. IEEE Trans. Comput. **64**(3), 818–831 (2015)
22. M. Clerc, J. Kennedy, The particle swarm-explosion, stability, and convergence in a multidimensional complex space. IEEE Trans. Evol. Comput. **6**(1), 58–73 (2002)
23. S. Ioffe, C. Szegedy, Batch normalization: accelerating deep network training by reducing internal covariate shift, in *Proceedings of Proceedings of the 32nd International Conference on International Conference on Machine Learning (ICML)* vol. 37 (2015), pp. 448–456
24. Didi Chuxing GAIA Initiative. https://gaia.didichuxing.com. Accessed July 2021

Chapter 5
Load- and Mobility-Aware Cooperative Content Delivery in the SAGVN

Abstract In this chapter, we investigate cooperative content delivery in the mobile edge caching-assisted space–air–ground integrated vehicular networks (SAGVNs), where vehicular content requests can be simultaneously served by multiple APs in space, aerial, and terrestrial networks. To minimize the overall content delivery delay while satisfying vehicular quality-of-service (QoS) requirements, we formulate a joint optimization problem of vehicle-to-AP *a*ssociation, *b*andwidth allocation, and *c*ontent delivery ratio, referred to as the *ABC* problem. To address the tightly coupled optimization variables, we propose a load- and mobility-aware *ABC* (*LMA-ABC*) scheme to solve the joint optimization problem. Specifically, we first decompose the *ABC* problem to optimize the content delivery ratio. Then the impact of bandwidth allocation on the achievable delay performance is analyzed, and an effect of diminishing delay performance gain is revealed. Based on the analysis results, the *LMA-ABC* scheme is designed with the consideration of user fairness, load balancing, and vehicle mobility. Simulation results demonstrate that the proposed *LMA-ABC* scheme can significantly reduce the cooperative content delivery delay comparing to the benchmark schemes.

To support multifarious vehicular services with diversified QoS requirements, the SAGVN is envisioned as a promising paradigm to provide global network coverage, enhance network flexibility, and improve network reliability [1–3]. In the SAGVN, CAVs should be able to access their requested data with minimal latency. To achieve this goal, different network segments in the SAGVN should cooperate in vehicular service provisioning by leveraging heterogeneous network resources ingeniously. With cooperative content delivery, CAVs can be served by multiple APs (including terrestrial BSs, aerial UAVs, and space satellites) simultaneously to enhance the service quality. However, achieving efficient cooperative delivery is a daunting task in the SAGVN, since significant research issues arise, including (1) vehicle-to-AP association, (2) wireless communication resource allocation, and (3) content delivery ratio optimization (i.e., determining the content delivery ratio at different APs), which are crucial to content delivery performance yet hard to be addressed due to their intercoupling relationships.

In the literature, user association and resource allocation have been widely studied. In [4], a network-wide user association problem is studied in heterogeneous cellular networks, where a low-complexity algorithm is proposed to find a near-optimal solution. A joint user association and resource allocation problem is investigated in [5] and solved by employing decomposition methods. In [6], a problem of joint user association, channel allocation, antenna selection, and power control is studied. Recently, RL-based methods have attracted growing research attention to solving user association and resource allocation problems [7–9]. However, these studies focus on single-AP association, where each user can be associated with at most one AP each time. This significantly constrains the communication performance including throughput and reliability [10]. In this chapter, we focus on a more complicated cooperative content delivery scenario with multi-AP association to enhance vehicular service quality, where the existing solutions cannot be applied. Furthermore, distinct characteristics of different network segments should be considered when making decisions on the vehicle-to-AP association, bandwidth allocation, and delivery ratio optimization, which are not considered in the above-mentioned existing works.

In this chapter, we focus on cooperative content delivery in the SAGVN, where LEO satellites, UAVs, and ground BSs cooperatively serve the CAV content requests. Specifically, an *ABC* problem is formulated to jointly optimize the vehicle-to-AP association, bandwidth allocation, and content delivery ratio, with the objective of minimizing the overall content delivery delay while satisfying the QoS requirements. The joint optimization problem is intractable due to the following reasons. First, the vehicle-to-AP association is tricky in the complicated SAGVN scenario due to the unprecedented heterogeneity and large network scale. Second, the wireless communication resource allocation should be optimized for all the service requests to guarantee the overall network performance, with user mobility and load balancing taken into consideration. Third, the content delivery ratio should be optimized considering differentiated network characteristics including network capacity, propagation delay, traffic load, and so on. Furthermore, these decisions are tightly coupled and involve both integer and continuous variables. To tackle these challenges, we propose an *LMA-ABC* scheme to solve the joint optimization problem. Specifically, we first decompose the *ABC* problem to optimize the content delivery ratio by considering differentiated network characteristics including network capacity and propagation delay. For the bandwidth resource allocation, a diminishing gain effect is revealed, i.e., with more bandwidth resources allocated to the same user, the performance gain in terms of content delivery delay becomes marginal. Based on the delivery ratio optimization and the diminishing gain effect, the *LMA-ABC* scheme is designed to solve the joint optimization problem with the consideration of user fairness, load balancing, and vehicle mobility. The main contributions are summarized as follows:

- We study the cooperative vehicular content delivery in the SAGVN to enable ingenious cooperation among different network segments. Specifically, we formulate the *ABC* problem, which jointly optimizes user association, spectrum resource allocation, and content delivery ratio, to reduce the overall content delivery delay, which is of paramount importance for CAV services.

- To efficiently utilize the heterogeneous network resources, we propose an *LMA-ABC* scheme to solve the *ABC* problem, where the impact of different variables on the achievable delay performance is analyzed by problem decomposition. By leveraging the interplay between vehicle mobility and heterogeneous network characteristics (e.g., network capacity and propagation delay), the *LMA-ABC* scheme can effectively solve the joint optimization problem to achieve user fairness, load balancing, and vehicle mobility.
- Extensive experimental results are conducted, and results show that the proposed *LMA-ABC* scheme can significantly reduce the overall content delivery delay comparing to the benchmark schemes.

The remainder of this chapter is organized as follows. Section 5.1 shows the system model and problem formulation. Section 5.2 provides the problem analysis to reveal the impact of different variables on the achievable content delivery delay. The proposed *LMA-ABC* scheme design is demonstrated in Sect. 5.3. The performance evaluation is carried out in Sect. 5.4, followed by a brief summary in Sect. 5.5.

5.1 System Model and Problem Formulation

5.1.1 Scenario Description and Assumptions

We consider an SAGVN scenario with N_S LEO satellites, N_U UAVs, N_B LTE BSs, and N_V vehicles, as shown in Fig. 5.1. Let $\mathcal{SAT} = \{s_1, \ldots, s_{N_S}\}$, $\mathcal{UAV} = \{u_1, \ldots, u_{N_U}\}$, $\mathcal{BS} = \{b_1, \ldots, b_{N_B}\}$, and $\mathcal{V} = \{v_1, \ldots, v_{N_V}\}$ denote the set of LEO satellites, UAVs, BSs, and vehicles, respectively. $\mathcal{AP}_{all} = \mathcal{SAT} \cup \mathcal{UAV} \cup \mathcal{BS}$ denotes the set of all the APs. Each vehicular user is equipped with three radio interfaces for LTE, UAV, and satellite communications, respectively. Notice that satellites, UAVs, and BSs use orthogonal spectrum frequencies for communications to avoid co-channel interference. The available spectrum bandwidth for AP ap is B_{ap}, $ap \in \mathcal{AP}_{all}$. In this chapter, we adopt the control architecture proposed in [1], where the BSs, UAVs, and satellites are controlled by a centralized controller to perform cooperative content delivery management.

In this chapter, ground BSs, UAVs, and satellites are caching-enabled, i.e., some content files have already been cached in the APs to serve the vehicles. Therefore, the unstable and capacity-limited UAV backhaul and the satellite feeder links can be avoided. Considering that satellites can cover users all around the world, caching commonly popular files (e.g., pop music, hot movies, global news, etc.) on satellites can achieve a good caching performance gain. On the other hand, files cached in UAVs can be determined based on the flying trajectories and user request profiles [11, 12], and the BSs can cache files based on the BS–vehicle contact duration, file

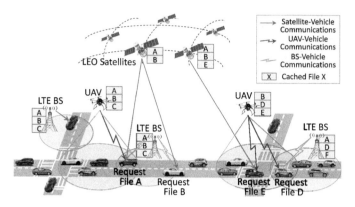

Fig. 5.1 Illustration of cooperative content delivery in SAGVN

request pattern, and caching storage capacities [13, 14]. Denote by $\mathcal{F}$ the file library containing all the F content files, and the size of file f is denoted by ς_f, $\forall f \in \mathcal{F}$. Denote by $c_{ap,f}$ the caching indicator, $c_{ap,f} = 1$ if file f is cached in AP ap, otherwise $c_{ap,f} = 0$, $\forall ap \in \mathcal{AP}_{all}$, $\forall f \in \mathcal{F}$.

5.1.2 *Communication Model*

In the SAGVN, vehicular content requests can be served by different types of APs simultaneously. When vehicle v requests content file f, the satellites, UAVs, and BSs that have cached file f can act as the candidate APs. Let $\delta_{v,f}$ denote the content request indicator, $\delta_{v,f} = 1$ when vehicle v requests for content f, and $\delta_{v,f} = 0$ otherwise. Without loss of generality, each vehicle requests at most one content file at each time [15]. Considering the mobility of vehicles, UAVs, and LEO satellites, the remaining contact time between vehicle v and AP ap is denoted by $T_{ap,v}^{\text{rem}}$.[1] When vehicle v requests file f, the set of candidate APs is

$$\mathcal{AP}_{v,f} = \left\{ ap \mid ap \in \mathcal{AP}_{all}, \delta_{v,f} \cdot c_{ap,f} = 1, T_{ap,v}^{\text{rem}} > 0 \right\}. \tag{5.1}$$

Let $a_{ap,v}$ be the association indicator, $a_{ap,v} = 1$ when vehicle v is associated with AP ap, and $a_{ap,v} = 0$ otherwise. When vehicle v is within the coverage of multiple APs of the same type (e.g., multiple LTE BSs, UAVs, or satellites), it can connect to at most one AP from the same network segment at each time, i.e.,

---

[1] Vehicles can upload their locations to the centralized controller, based on which $T_{ap,v}^{\text{rem}}$ can be calculated with the fixed deployment of BSs and the trackable locations of UAVs and satellites.

$$\sum_{s_i \in \mathcal{SAT}} a_{s_i,v} \leq 1, \quad \sum_{u_j \in \mathcal{UAV}} a_{u_j,v} \leq 1, \quad \sum_{b_k \in \mathcal{BS}} a_{b_k,v} \leq 1. \tag{5.2}$$

Let $\varsigma_{ap,v,f}$ be the size of file f delivered from ap ($ap \in \mathcal{AP}_{v,f}$) to vehicle v ($v \in \mathcal{V}$). Thus we have

$$0 \leq \varsigma_{ap,v,f} \leq \varsigma_f, \quad \sum_{AP \in \mathcal{AP}_{v,f}} \varsigma_{ap,v,f} = \varsigma_f. \tag{5.3}$$

BS-to-Vehicle (B2V) Communications For a B2V communication link between BS b_k and vehicle v, the achievable signal-to-noise ratio (SNR) is derived as

$$\Gamma_{b_k,v} = P_{b_k,v} d_{b_k,v}^{-\alpha} h_{b_k,v} / \sigma^2, \tag{5.4}$$

where $P_{b_k,v}$, $d_{b_k,v}$, $h_{b_k,v}$, α, and σ^2 are the transmit power of BS b_k, the distance between b_k and vehicle v, the channel fading (following Rayleigh fading) from b_k to v, the pathloss exponent, and the Gaussian noise power. With an allocated spectrum bandwidth of $B_{b_k,v}$ from BS b_k, the achievable data rate of B2V communication is $R_{b_k,v} = B_{b_k,v} \log_2(1 + \Gamma_{b_k,v})$.

UAV-to-Vehicle (U2V) Communications The achievable SNR of a U2V link is

$$\Gamma_{u_j,v} = P_{u_j,v} PL_{u_j,v} h_{u_j,v} / \sigma^2, \tag{5.5}$$

where $P_{u_j,v}$ is the transmit power of UAV u_j, $PL_{u_j,v}$ is the pathloss from u_j to vehicle v consisting of LoS and NLOS components [11], and $h_{u_j,v}$ is the Rayleigh channel fading. With an allocated bandwidth of $B_{u_j,v}$ from UAV u_j, the achievable U2V data rate is $R_{u_j,v} = B_{u_j,v} \log_2(1 + \Gamma_{u_j,v})$.

Satellite-to-Vehicle (S2V) Communications The achievable SNR of an S2V link between satellite s_i and vehicle v is

$$\Gamma_{s_i,v} = P_{s_i,v} d_{s_i,v}^{-\alpha} h_{s_i,v} / \sigma^2, \tag{5.6}$$

where $P_{s_i,v}$ is the transmit power of satellite s_i, $d_{s_i,v}$ is distance between s_i and vehicle v, and $h_{s_i,v}$ is the channel fading. For S2V communications, the LoS signal is a strong dominant component. Thus, the S2V channels are considered as Rician fading channels [16], with the probability density function of the channel fading as

$$f(h) = \frac{K+1}{\Omega} \exp\left\{-K - \frac{(K+1)h}{\Omega}\right\} I_0\left(2\sqrt{\frac{K(K+1)h}{\Omega}}\right), \tag{5.7}$$

where K is the ratio between the power in the LoS path and the power in the other scattered paths, Ω is the total power of the LOS and scattering signals, and $I_0(\cdot)$ is the modified Bessel function of the first kind with zero order [16]. With an allocated bandwidth of $B_{s_i,v}$ from satellite s_i, the achievable S2V data rate is $R_{s_i,v} = B_{s_i,v} \log_2(1 + \Gamma_{s_i,v})$.

For B2V and U2V content delivery, the propagation delay is negligible. However, the propagation delay for satellite communications is not negligible due to the long communication distance. Considering the trackability of satellites, the position and the elevation angle of satellites are available at any time. For a given observation point (e.g., vehicle v), the altitude and elevation angle of LEO satellite s_i can be observed and denoted by h_{s_i} and ε_{s_i}, respectively. Thus, the S2V communication distance $d_{s_i,v}$ can be calculated as

$$r_e^2 + d_{s_i,v}^2 - 2r_e d_{s_i,v} \cos\left(\varepsilon_{s_i} + \frac{\pi}{2}\right) = (r_e + h_{s_i})^2 \tag{5.8}$$

$$\Rightarrow d_{s_i,v} = \sqrt{h_{s_i}^2 + 2h_{s_i}r_e + r_e^2 \sin^2 \varepsilon_{s_i}} - r_e \sin \varepsilon_{s_i},$$

where r_e is the earth radius. Thus, the propagation delay from s_i to v is $D_{s_i,v}^{\text{prop}} = d_{s_i,v}/c$, where c is the speed of light.

5.1.3 Problem Formulation

To minimize the overall content delivery delay for all content requests, we investigate the *ABC* problem to jointly optimize (1) vehicle-to-AP association $a = \{a_{ap,v}\}$, (2) bandwidth allocation $b = \{B_{ap,v}\}$, and (3) content delivery ratio $\varsigma = \{\varsigma_{ap,v,f}\}$. Notice that, when a vehicle is associated with multiple APs for cooperative content delivery, the overall delivery delay is determined by the worst-case link, i.e., the link with the longest delivery delay. Therefore, with cooperative delivery decisions a, b, and ς, the expected delay of delivering file f to vehicle v is

$$D_{v,f} = \max\left\{ \sum_{s_i \in \mathcal{SAT}} a_{s_i,v}\left(\frac{\varsigma_{s_i,v,f}}{R_{s_i,v}} + D_{s_i,v}^{\text{prop}}\right), \sum_{u_j \in \mathcal{UAV}} \frac{a_{u_j,v}\varsigma_{u_j,v,f}}{R_{u_j,v}}, \right. \tag{5.9}$$

$$\left. \sum_{b_k \in \mathcal{BS}} \frac{a_{b_k,v}\varsigma_{b_k,v,f}}{R_{b_k,v}} \right\} + \left(1 - \max_{ap \in \mathcal{AP}_{v,f}}\{a_{ap,v}\}\right) D_{\max}.$$

The last term in (5.9) indicates that, when vehicle v is not associated with any APs, the content delivery fails and the corresponding content delivery delay is $D_{\max}$, which is a sufficiently large number to penalize the unsuccessful content delivery. To this end, the *ABC* problem can be formulated as

$$\min_{a, b, \varsigma} \sum_{v \in \mathcal{V}} \sum_{f \in \mathcal{F}} \delta_{v,f} D_{v,f} \tag{5.10}$$

$$\text{s.t.}\quad a_{ap,v} \in \{0, 1\},\quad \varsigma_{ap,v,f} \in [0, \varsigma_f], \tag{5.10a}$$

$$\sum_{s_i \in \mathcal{SAT}} a_{s_i,v} \le 1, \sum_{u_j \in \mathcal{UAV}} a_{u_j,v} \le 1, \tag{5.10b}$$

$$\sum_{b_k \in BS} a_{b_k,v} \leq 1, \sum_{ap \in \mathcal{AP}_{v,f}} \varsigma_{ap,v,f} = \varsigma_f, \tag{5.10c}$$

$$0 \leq B_{ap,v} \leq B_{ap}, \quad \sum_{v \in V} B_{ap,v} \leq B_{ap}, \tag{5.10d}$$

$$a_{ap,v} \leq \mathbb{1}_{T_{ap,v}^{rem}>0} \, c_{ap,f} \, \delta_{v,f}, \tag{5.10e}$$

$$a_{ap,v} \Gamma_{ap,v} \geq a_{ap,v} \Gamma_{th}, \quad \forall ap \in \mathcal{AP}_{all}, \forall v \in V, \tag{5.10f}$$

where Γ_{th} is the minimum SNR requirement for correct data detection at the receiving vehicle. $\mathbb{1}_{condition}$ is an indicator, where $\mathbb{1}_{condition} = 1$ if the *condition* is true, and $\mathbb{1}_{condition} = 0$ otherwise. Constraint (5.10d) means that the allocated bandwidth cannot exceed the total available bandwidth resources. Constraints (5.10e) and (5.10f) indicate that vehicle v can be associated with an AP only when the AP belongs to the candidate set $\mathcal{AP}_{v,f}$ and $\Gamma_{ap,v} \geq \Gamma_{th}$.

5.2   Problem Analysis Based on Decomposition

The optimization problem (5.10) is non-convex and intractable since it involves both integer and continuous variables, and decision variables a, b, and ς are tightly coupled. To solve this problem, we first decompose the *ABC* problem to analyze the impact of content delivery ratio and bandwidth allocation on the achievable content delivery delay performance in this section.

5.2.1   Optimization of ς with Known a and b

When given the AP–vehicle association and the corresponding bandwidth allocation, each vehicle v can easily calculate the achievable data rate from different APs. The associated satellite, UAV, and BS and the corresponding achievable data rates for vehicle v can be, respectively, expressed as

$$s_i^* = \{s_i \in \mathcal{SAT} | a_{s_i,v} = 1\}, \quad R_{s_i^*,v} = B_{s_i^*,v} \log_2(1 + \Gamma_{s_i^*,v}), \tag{5.11}$$

$$u_j^* = \{u_j \in \mathcal{UAV} | a_{u_j,v} = 1\}, \quad R_{u_j^*,v} = B_{u_j^*,v} \log_2(1 + \Gamma_{u_j^*,v}),$$

$$b_k^* = \{b_k \in \mathcal{BS} | a_{b_k,v} = 1\}, \quad R_{b_k^*,v} = B_{b_k^*,v} \log_2(1 + \Gamma_{b_k^*,v}).$$

If vehicle v is not associated with any satellites, UAVs, or BSs, we have $s_i^* = \emptyset$, $u_j^* = \emptyset$, or $b_k^* = \emptyset$. Notice that, for the case with $s_i^* = u_j^* = b_k^* = \emptyset$ or $R_{s_i^*,v} + R_{u_j^*,v} + R_{b_k^*,v} = 0$, the optimal delivery ratio is $\varsigma_{s_i^*,v} = \varsigma_{u_j^*,v} = \varsigma_{b_k^*,v} = 0$, and the corresponding content delivery delay is $D_{v,f} = D_{\max}$. For notational simplicity, we use $\mathbb{1}_{s_i^*}$, $\mathbb{1}_{u_j^*}$, and $\mathbb{1}_{b_k^*}$ to represent $\mathbb{1}_{s_i^* \neq \emptyset}$, $\mathbb{1}_{u_j^* \neq \emptyset}$, and $\mathbb{1}_{b_k^* \neq \emptyset}$, respectively. The optimal content delivery ratio for vehicle v can be derived as given in the following lemma, considering the non-trivial case where $\mathbb{1}_{s_i^*} + \mathbb{1}_{u_j^*} + \mathbb{1}_{b_k^*} \geq 1$.

Lemma 5.1 *With known $\boldsymbol{a}$, $\boldsymbol{b}$, and the corresponding achievable data rates, the optimal content delivery ratio is*

$$
\varsigma_{s_i^*,v,f} = \frac{\mathbb{1}_{s_i^*} R_{s_i^*,v} \max\left\{ \varsigma_f - D_{s_i^*,v}^{\mathrm{prop}} \left(\mathbb{1}_{u_j^*} R_{u_j^*,v} + \mathbb{1}_{b_k^*} R_{b_k^*,v} \right), 0 \right\}}{R_{s_i^*,v} + \mathbb{1}_{b_k^*} R_{b_k^*,v} + \mathbb{1}_{u_j^*} R_{u_j^*,v}},
$$

$$
\varsigma_{b_k^*,v,f} = \frac{\mathbb{1}_{b_k^*} R_{b_k^*,v} \left[\varsigma_f + \mathbb{1}_{s_i^*} \mathbb{1}_{\frac{\varsigma_f}{\mathbb{1}_{u_j^*} R_{u_j^*,v} + R_{b_k^*,v}} > D_{s_i^*,v}^{\mathrm{prop}}} D_{s_i^*,v}^{\mathrm{prop}} R_{s_i^*,v} \right]}{\mathbb{1}_{s_i^*} \mathbb{1}_{\frac{\varsigma_f}{\mathbb{1}_{u_j^*} R_{u_j^*,v} + R_{b_k^*,v}} > D_{s_i^*,v}^{\mathrm{prop}}} R_{s_i^*,v} + R_{b_k^*,v} + \mathbb{1}_{u_j^*} R_{u_j^*,v}},
$$

$$
\varsigma_{u_j^*,v,f} = \frac{\mathbb{1}_{u_j^*} R_{u_j^*,v} \left[\varsigma_f + \mathbb{1}_{s_i^*} \mathbb{1}_{\frac{\varsigma_f}{R_{u_j^*,v} + \mathbb{1}_{b_k^*} R_{b_k^*,v}} > D_{s_i^*,v}^{\mathrm{prop}}} D_{s_i^*,v}^{\mathrm{prop}} R_{s_i^*,v} \right]}{\mathbb{1}_{s_i^*} \mathbb{1}_{\frac{\varsigma_f}{R_{u_j^*,v} + \mathbb{1}_{b_k^*} R_{b_k^*,v}} > D_{s_i^*,v}^{\mathrm{prop}}} R_{s_i^*,v} + \mathbb{1}_{b_k^*} R_{b_k^*,v} + R_{u_j^*,v}}.
$$

Proof Considering the non-negligible satellite propagation delay, in the following, the content delivery ratio is optimized by considering two cases based on whether satellites participate in the cooperative content delivery.

Case 1: $\mathbb{1}_{s_i^*} = 0$ **or** $\frac{\varsigma_f}{\mathbb{1}_{u_j^*} R_{u_j^*,v} + \mathbb{1}_{b_k^*} R_{b_k^*,v}} \leq D_{s_i^*,v}^{\mathrm{prop}}$: When $s_i^* = \varnothing$ or the content delivery can be accomplished by the BS and/or the UAV with a delay shorter than the satellite propagation delay $D_{s_i^*,v}^{\mathrm{prop}}$, the satellite does not participate in the content delivery, and the optimal solution is

$$
\Rightarrow
\begin{cases}
\varsigma_{u_j^*,v,f} = \mathbb{1}_{u_j^*} \varsigma_f, \quad \varsigma_{b_k^*,v,f} = \mathbb{1}_{b_k^*} \varsigma_f, \quad \text{if } \mathbb{1}_{u_j^*} + \mathbb{1}_{b_k^*} = 1, \\[2mm]
\dfrac{\varsigma_{u_j^*,v,f}}{R_{u_j^*,v}} = \dfrac{\varsigma_{b_k^*,v,f}}{R_{b_k^*,v}}, \quad\quad\quad \text{if } \mathbb{1}_{u_j^*} + \mathbb{1}_{b_k^*} = 2, \\[4mm]
\varsigma_{u_j^*,v,f} = \dfrac{\mathbb{1}_{u_j^*} R_{u_j^*,v} \varsigma_f}{\mathbb{1}_{u_j^*} R_{u_j^*,v} + \mathbb{1}_{b_k^*} R_{b_k^*,v}}, \\[3mm]
\varsigma_{b_k^*,v,f} = \dfrac{\mathbb{1}_{b_k^*} R_{b_k^*,v} \varsigma_f}{\mathbb{1}_{u_j^*} R_{u_j^*,v} + \mathbb{1}_{b_k^*} R_{b_k^*,v}}, \\[3mm]
\varsigma_{s_i^*,v,f} = 0.
\end{cases}
$$

Case 2: $\mathbb{1}_{s_i^*} = 1$ **and** $\frac{\varsigma_f}{\mathbb{1}_{u_j^*} R_{u_j^*,v} + \mathbb{1}_{b_k^*} R_{b_k^*,v}} > D_{s_i^*,v}^{\mathrm{prop}}$: When the content delivery accomplished by the BS and/or the UAV has a delay longer than $D_{s_i^*,v}^{\mathrm{prop}}$, the satellite can participate in content delivery to further reduce the delay. In this case, we have

$$
\begin{cases}
\varsigma_{s_i^*,v,f} = \varsigma_f, \quad \varsigma_{u_j^*,v,f} = \varsigma_{b_k^*,v,f} = 0, & \text{if } \mathbb{1}_{u_j^*} + \mathbb{1}_{b_k^*} = 0, \\[2mm]
\dfrac{\varsigma_{s_i^*,v,f}}{R_{s_i^*,v}} + D_{s_i^*,v}^{\mathrm{prop}} = \mathbb{1}_{u_j^*}\dfrac{\varsigma_{u_j^*,v,f}}{R_{u_j^*,v}} + \mathbb{1}_{b_k^*}\dfrac{\varsigma_{b_k^*,v,f}}{R_{b_k^*,v}}, & \text{if } \mathbb{1}_{u_j^*} + \mathbb{1}_{b_k^*} = 1, \\[2mm]
\dfrac{\varsigma_{s_i^*,v,f}}{R_{s_i^*,v}} + D_{s_i^*,v}^{\mathrm{prop}} = \dfrac{\varsigma_{u_j^*,v,f}}{R_{u_j^*,v}} = \dfrac{\varsigma_{b_k^*,v,f}}{R_{b_k^*,v}}, & \text{if } \mathbb{1}_{u_j^*} + \mathbb{1}_{b_k^*} = 2,
\end{cases}
$$

$$
\Rightarrow
\begin{cases}
\varsigma_{s_i^*,v,f} = \dfrac{R_{s_i^*,v}\varsigma_f - D_{s_i^*,v}^{\mathrm{prop}} R_{s_i^*,v}\left(\mathbb{1}_{u_j^*} R_{u_j^*,v} + \mathbb{1}_{b_k^*} R_{b_k^*,v}\right)}{R_{s_i^*,v} + \mathbb{1}_{b_k^*} R_{b_k^*,v} + \mathbb{1}_{u_j^*} R_{u_j^*,v}}, \\[4mm]
\varsigma_{b_k^*,v,f} = \mathbb{1}_{b_k^*}\dfrac{R_{b_k^*,v}\cdot\varsigma_f + D_{s_i^*,v}^{\mathrm{prop}} R_{s_i^*,v} R_{b_k^*,v}}{R_{s_i^*,v} + \mathbb{1}_{b_k^*} R_{b_k^*,v} + \mathbb{1}_{u_j^*} R_{u_j^*,v}}, \\[4mm]
\varsigma_{u_j^*,v,f} = \mathbb{1}_{u_j^*}\dfrac{R_{u_j^*,v}\cdot\varsigma_f + D_{s_i^*,v}^{\mathrm{prop}} R_{s_i^*,v} R_{u_j^*,v}}{R_{s_i^*,v} + \mathbb{1}_{b_k^*} R_{b_k^*,v} + \mathbb{1}_{u_j^*} R_{u_j^*,v}}.
\end{cases}
$$

Combining the above two cases, we can derive the optimal content delivery ratio as expressed in Lemma 5.1. □

5.2.2   Delay Performance Gain with Bandwidth Allocation

In this part, we investigate the impact of $\boldsymbol{b}$ on the achievable content delivery delay, when given $\boldsymbol{a}$.

Lemma 5.2 (Diminishing Gain Effect) *For bandwidth allocation in each AP (i.e., BS, UAV, or satellite), with more bandwidth resources allocated to the same vehicular user, the delay performance gain (i.e., delay decrement) diminishes.*

Proof Let $\boldsymbol{b}_0$ be the initial bandwidth allocation policy. The optimal content delivery ratio and the corresponding overall content delivery delay performance can be calculated based on Lemma 5.1 and (5.9). Taking B2V communications as an example, for vehicle v that is connected with BS b_k^* for the given $\boldsymbol{a}$ and $\boldsymbol{b}_0$, its content retrieving delay is

$$
D_{v,f}(\boldsymbol{a},\boldsymbol{b}_0) = \frac{\varsigma_{b_k^*,v,f}}{R_{b_k^*,v}} = \frac{\varsigma_f + \mathbb{1}_{s_i^*}\mathbb{1}\frac{\varsigma_f}{\mathbb{1}_{u_j^*} R_{u_j^*,v} + R_{b_k^*,v}} > D_{s_i^*,v}^{\mathrm{prop}} \, D_{s_i^*,v}^{\mathrm{prop}} R_{s_i^*,v}}{\mathbb{1}_{s_i^*}\mathbb{1}\frac{\varsigma_f}{\mathbb{1}_{u_j^*} R_{u_j^*,v} + R_{b_k^*,v}} > D_{s_i^*,v}^{\mathrm{prop}} R_{s_i^*,v} + B_{b_k^*,v}^0 \gamma_{b_k^*,v} + \mathbb{1}_{u_j^*} R_{u_j^*,v}}, \tag{5.12}
$$

where $B_{b_k^*,v}^0$ is the bandwidth allocated from BS b_k^* to vehicle v with the given $\boldsymbol{b}_0$, and $\gamma_{b_k^*,v} = \log_2(1 + \Gamma_{b_k^*,v})$ is the spectrum efficiency of the B2V communication.

For a new bandwidth allocation decision $\boldsymbol{b}'$, in which BS b_k^* allocates an extra bandwidth of $\Delta B_{b_k^*,v}$ to vehicle v (the other allocation decisions keep the same with $\boldsymbol{b}_0$), the delay of delivering content f to vehicle v is

$$D_{v,f}(\boldsymbol{a}, \boldsymbol{b}') = \frac{\varsigma_f + \mathbb{1}_{s_i^* \mathbb{1}_{\frac{\varsigma_f}{\mathbb{1}_{u_j^*} R_{u_j^*,v} + R_{b_k^*,v}} > D_{s_i^*,v}^{\mathrm{prop}}} D_{s_i^*,v}^{\mathrm{prop}} R_{s_i^*,v}}}{\mathbb{1}_{s_i^* \mathbb{1}_{\frac{\varsigma_f}{\mathbb{1}_{u_j^*} R_{u_j^*,v} + R_{b_k^*,v}} > D_{s_i^*,v}^{\mathrm{prop}}} R_{s_i^*,v} + (B_{b_k^*,v}^0 + \Delta B_{b_k^*,v})\gamma_{b_k^*,v} + \mathbb{1}_{u_j^*} R_{u_j^*,v}}. \tag{5.13}$$

When the value of $\mathbb{1}_{\frac{\varsigma_f}{\mathbb{1}_{u_j^*} R_{u_j^*,v} + R_{b_k^*,v}} > D_{s_i^*,v}^{\mathrm{prop}}}$ keeps unchanged for decisions $\boldsymbol{b}$ and $\boldsymbol{b}'$, for notational simplicity, let

$$\mu = \varsigma_f + \mathbb{1}_{s_i^* \mathbb{1}_{\frac{\varsigma_f}{\mathbb{1}_{u_j^*} R_{u_j^*,v} + R_{b_k^*,v}} > D_{s_i^*,v}^{\mathrm{prop}}}} D_{s_i^*,v}^{\mathrm{prop}} R_{s_i^*,v},$$

$$\nu = \mathbb{1}_{s_i^* \mathbb{1}_{\frac{\varsigma_f}{\mathbb{1}_{u_j^*} R_{u_j^*,v} + R_{b_k^*,v}} > D_{s_i^*,v}^{\mathrm{prop}}}} R_{s_i^*,v} + B_{b_k^*,v}^0 \gamma_{b_k^*,v} + \mathbb{1}_{u_j^*} R_{u_j^*,v},$$

and then the delay performance gain (i.e., delay decrement) is

$$\Delta D_{v,f}(\Delta B_{b_k^*,v}) = D_{v,f}(\boldsymbol{a}, \boldsymbol{b}_0) - D_{v,f}(\boldsymbol{a}, \boldsymbol{b}') = \frac{\mu \gamma_{b_k^*,v} \Delta B_{b_k^*,v}}{\nu(\nu + \Delta B_{b_k^*,v} \gamma_{b_k^*,v})}. \tag{5.14}$$

If the value of $\mathbb{1}_{\frac{\varsigma_f}{\mathbb{1}_{u_j^*} R_{u_j^*,v} + R_{b_k^*,v}} > D_{s_i^*,v}^{\mathrm{prop}}}$ changes, i.e., $\mathbb{1}_{\frac{\varsigma_f}{\mathbb{1}_{u_j^*} R_{u_j^*,v} + R_{b_k^*,v}} > D_{s_i^*,v}^{\mathrm{prop}}} = 1$ for $\boldsymbol{b}$ and $\mathbb{1}_{\frac{\varsigma_f}{\mathbb{1}_{u_j^*} R_{u_j^*,v} + R_{b_k^*,v}} > D_{s_i^*,v}^{\mathrm{prop}}} = 0$ for $\boldsymbol{b}'$, the delay performance gain is

$$\Delta D_{v,f}(\Delta B_{b_k^*,v}) = \frac{\varsigma_f + \mathbb{1}_{s_i^*} D_{s_i^*,v}^{\mathrm{prop}} R_{s_i^*,v}}{\mathbb{1}_{s_i^*} R_{s_i^*,v} + B_{b_k^*,v}^0 \gamma_{b_k^*,v} + \mathbb{1}_{u_j^*} R_{u_j^*,v}} - \frac{\varsigma_f}{(B_{b_k^*,v}^0 + \Delta B_{b_k^*,v})\gamma_{b_k^*,v} + \mathbb{1}_{u_j^*} R_{u_j^*,v}}$$

$$= \frac{\Delta B_{b_k^*,v} \gamma_{b_k^*,v}(\varsigma_f + \mathbb{1}_{s_i^*} D_{s_i^*,v}^{\mathrm{prop}} R_{s_i^*,v}) - \mathbb{1}_{s_i^*} R_{s_i^*,v}\left[\varsigma_f - D_{s_i^*,v}^{\mathrm{prop}}(B_{b_k^*,v}^0 \gamma_{b_k^*,v} + \mathbb{1}_{u_j^*} R_{u_j^*,v})\right]}{\left[\mathbb{1}_{s_i^*} R_{s_i^*,v} + B_{b_k^*,v}^0 \gamma_{b_k^*,v} + \mathbb{1}_{u_j^*} R_{u_j^*,v}\right]\left[(B_{b_k^*,v}^0 + \Delta B_{b_k^*,v})\gamma_{b_k^*,v} + \mathbb{1}_{u_j^*} R_{u_j^*,v}\right]}. \tag{5.15}$$

For the above two cases, the second derivative of $\Delta B_{b_k^*,v}$ is negative for each of them, which means that the delay performance gain is a concave function of $\Delta B_{b_k^*,v}$. In other words, when BS b_k^* allocates more bandwidth to vehicle v, the delay performance gain diminishes. Similarly, when considering bandwidth allocation from each UAV/satellite to a vehicle, the diminishing gain effect also exists given that the others' allocation decisions are fixed, which can conclude the proof. □

The diminishing gain effect naturally promotes user fairness in our scheme design. This is consistent with the resource allocation in real systems, where improving the well-served users' performance generally has a lower priority than allocating resources to users with unsatisfactory performance.

5.3   LMA-ABC Scheme for Cooperative Content Delivery

Based on the analysis presented in Sect. 5.2, we propose an *LMA-ABC* scheme to solve the *ABC* problem by taking user fairness, load balancing, and vehicle mobility into consideration.

5.3.1   Posterior Association Determination

In the *ABC* problem, the vehicle-to-AP association a can significantly affect the optimization of ς and b, as analyzed in Sects. 5.2.1 and 5.2.2. On the other hand, the optimization results of ς and b can also affect the determination of a:

- For a given a with $a_{ap,v} = 1$, if the bandwidth allocation decision is $B_{ap,v} = 0$ to minimize the overall delay, the association should be adjusted to avoid unnecessary association cost.
- Based on Lemma 5.1, there exists a special case with $a_{s_i^*,v} = 1$, $\varsigma_{s_i^*,v,f} = 0$ due to the long satellite propagation delay. In this case, the corresponding $a_{s_i^*,v}$ (and also $B_{s_i^*,v}$) should be adjusted to 0 to avoid undesirable resource waste.

Therefore, in the proposed scheme, to avoid inappropriate association decisions, we perform posterior association determination to decide the vehicle-to-AP association based on the optimization results of b and ς, i.e.,

$$a_{ap,v} = \mathbb{1}_{B_{ap,v}>0} \cdot \mathbb{1}_{\varsigma_{ap,v,f}>0}, \quad \forall ap \in \mathcal{AP}_{v,f}, \forall v \in \mathcal{V}. \tag{5.16}$$

5.3.2   Bandwidth Allocation with Diminishing Gain Effect

According to the diminishing gain effect in Lemma 5.2, it is undesirable to allocate a lot of bandwidth resources to the same user. Considering that the bandwidth resources for each AP consist of multiple sub-channels, in the proposed *LMA-ABC* scheme, we implement bandwidth allocation in the units of sub-channels. Specifically, we propose a greedy-based bandwidth allocation scheme considering the diminishing gain effect to improve the content delivery performance, detailed as follows:

- **Step 1:** According to the analysis in Lemma 5.2, the achievable delay performance gain for an AP's bandwidth allocation decision is affected by the other APs' decisions. In other words, making bandwidth allocation decisions for one AP can greatly impact the subsequent bandwidth allocation. To avoid prioritizing the APs (i.e., with different bandwidth allocation order), where different AP priorities might affect the final delay performance, in the proposed *LMA-ABC* scheme, all the resources are centrally managed by the central controller.

- **Step 2:** Based on vehicles' content requests ($\delta_{v,f}$), the cached content availability ($c_{ap,f}$), and link quality ($\Gamma_{ap,v}$), we can construct a connected graph between SAGVN APs and vehicles, where a v-to-ap connection is feasible if and only if $c_{ap,f}\, \delta_{ap,f}\, \mathbb{1}_{T_{ap,v}^{\mathrm{rem}}>0} = 1$ and $\Gamma_{ap,v} \geq \Gamma_{th}$.
- **Step 3:** For each feasible v-to-ap connection, when a sub-channel is allocated, the corresponding ς can be calculated based on Lemma 5.1 and the delay performance gain can be calculated based on (5.14), with the other APs' bandwidth allocation decisions unchanged. Therefore, the sub-channel–vehicle association can be selected with the best delay performance gain.
- **Step 4:** Recall that one vehicle can be associated with at most one AP of the same type (i.e., at most one BS, one UAV, and one satellite). Therefore, after performing **Step 3**, the connected graph should be updated by removing all the connections between the selected vehicle and the other APs of the same type. Repeat **Step 3** and **Step 4** until bandwidth resource depletion.

5.3.3   LMA-ABC Scheme Design

The bandwidth allocation scheme designed in Sect. 5.3.2 can effectively guarantee user fairness and enhance delivery performance, by preventing APs from allocating all the resources to a single user. However, the potential AP overloading problems might still occur. As shown in (5.14), the delay decrements are affected by not only the other APs' bandwidth allocation decisions but also the channel quality (represented by the spectrum efficiency $\gamma_{b_k,v}$). Taking B2V communications as an example, if there exists a BS b_k which is the best choice with the largest $\gamma_{b_k,v}$ for all the vehicles, then getting a sub-channel from BS b_k always has a higher delay performance gain than getting a sub-channel from other BSs. This may result in load imbalance among APs, where all the vehicles are associated with BS b_k, while the resources of other BSs are wasted. Another type of improper association might happen when associating vehicles to the APs with a small $T_{ap,v}^{\mathrm{rem}}$. With vehicle mobility, the association becomes invalid shortly. For vehicles located in multiple APs' coverage overlapping area, improper association can lead to unnecessary handover, which degrades the network performance. Furthermore, it may also lead to frequent re-execution of the *LMA-ABC* scheme and consume substantial computational resources. These problems are caused by the short-sighted gain calculation focusing only on the performance gain achieved by each sub-channel allocation.

To address these issues, we propose a far-sighted gain design in the *LMA-ABC* scheme by considering the potential future gain. When calculating the delay performance gain for a potential v-to-ap connection in **Step 3** in Sect. 5.3.2, if v has not been associated with any APs that have the same type with ap, we design a load- and mobility-aware gain as follows:

Algorithm 5: Procedure of the *LMA-ABC* scheme

Initialization:
$a_{ap,v} = 0$, $B_{ap,v} = 0$, and $\varsigma_{ap,v,f} = 0$, $\forall ap \in \mathcal{AP}_{all}$, $\forall v \in \mathcal{V}$.
Phase 1: Construct a connected graph $\mathcal{G}$ between the APs and the vehicles according to
Step 2 in Sect. 5.3.2.
Phase 2:
for *each sub-channel with bandwidth* ΔB_{ap}, $\forall ap \in \mathcal{AP}_{all}$ **do**
   | **for** $v \in \mathcal{V}$ *and v-to-ap connection is feasible in* $\mathcal{G}$ **do**
   | | Calculate the optimal ς based on Lemma 5.1.
   | | **if** *v has not been associated with any AP that has the same type with ap* **then**
   | | | Calculate the delay performance gain $\Delta D_{v,f,ap}^{all}$ based on (5.17).
   | | **else**
   | | | Calculate the delay performance gain $\Delta D_{v,f}(\Delta B_{ap,v})$ based on (5.14).
   | | **end**
   | **end**
end
Select the sub-channel–vehicle association with the largest delay performance gain.
Denote the selected AP and vehicle by ap^* and v^*, respectively.
Let $a_{ap^*,v^*} = 1$ and remove connections between v^* and the other APs that have the same
type with ap^* in $\mathcal{G}$.
Go to *Phase 2* and repeat until bandwidth resource depletes.
Phase 3: Calculate the optimal content delivery ratio ς based on the final results of a and b
according to Lemma 5.1.
Output: a, b, and ς.

$$\Delta D_{v,f,ap}^{all} = \Delta D_{v,f}(\Delta B_{ap}) + \lambda \Delta D_{v,f}\left(\frac{B_{ap}}{\sum_v a_{ap,v}}\right) + \beta T_{ap,v}^{rem}, \qquad (5.17)$$

where ΔB_{ap} is the bandwidth of a sub-channel allocated from ap, $\Delta D_{v,f}(\cdot)$ is the delay performance gain as defined in (5.14), T_{ap}^{rem} is the remaining contact time between v and ap, and $\lambda, \beta \in [0, 1]$ are constants. The second and third terms in the right part of (5.17) refer to the potential future gain that could be obtained when associating with ap, and parameters λ and β control the relative importance of the current gain and the future potential gain. With small λ and β, vehicles prefer to associate with the APs with good link quality, while for large λ and β, vehicles are more likely to associate with APs that are less crowded or have a long communication remaining time to achieve load balancing and avoid unnecessary handover overhead. The detailed procedure of the proposed *LMA-ABC* scheme for cooperative content delivery can be found in Algorithm 5.

5.4 Performance Evaluation

In this section, we conduct simulations to evaluate the performance of the proposed *LMA-ABC* scheme. The simulations are carried out based on the real scenario of University of Waterloo campus, where the VISSIM simulation tool is used to

Table 5.1 Simulation parameters

Number of LTE BSs	10
Number of UAVs	3
Number of LEO satellites	2
Altitude of satellites	781 km
Transmission power of LTE BSs $P_{b_k,v}$	28 dBm
Transmission power of UAVs	23 dBm
Transmission power of LEO satellites	10 dBW [17]
Satellite transmission antenna gain	20 dBi[17]
Satellite receiving antenna gain	30 ddBi[17]
Pathloss exponents of B2V communications	3.5
Pathloss exponents of U2V communications	2.8
Pathloss exponents of S2V communications	2.5
Available bandwidth for each LTE BS B_{b_k}	100 MHz
Available bandwidth for each UAV B_{u_j}	100 MHz
Available bandwidth for each LEO satellite B_{s_i}	500 MHz
Size of content files ς_f	[0 Mb, 100 Mb]
SNR threshold Γ_{th}	5 dB

emulate the vehicle traffic in the campus scenario. The simulation parameters are
summarized in Table 5.1.

Given that the content delivery delay is dominated by file sizes and the number
of requests, average delay per unit data (sec/Mbits) is used as a performance metric
in our simulation for fairness consideration. All the simulation results presented
in this section are averaged over 100 random testing trials. To further evaluate
the effectiveness of the *LMA-ABC* scheme, the following benchmark schemes are
considered for performance comparison:

- Best SNR association (BSA): All the vehicles are associated with the APs with
  the best SNR. The optimization of b and ς keeps the same as the *LMA-ABC*
  scheme.
- Random association (RA): All the vehicles are randomly associated with the APs.
  The optimization of b and ς keeps the same as the *LMA-ABC* scheme.
- Equal bandwidth allocation (EBA): Use Algorithm 5 to determine a. Bandwidth
  resources of each AP are equally allocated to all the associated vehicles, and ς is
  calculated based on Lemma 5.1.
- Equal throughput bandwidth allocation (ETBA): Use Algorithm 5 to determine
  a. Bandwidth resources of each AP are allocated such that all the associated
  vehicles have the same throughput, and ς is calculated based on Lemma 5.1.

In Fig. 5.2, we compare the traffic load balancing performance with different
association schemes. Without loss of generality, we use the standard deviation
(STD) of the number of vehicles associated with different APs to evaluate the
traffic load balancing performance. Generally, a smaller traffic load STD indicates

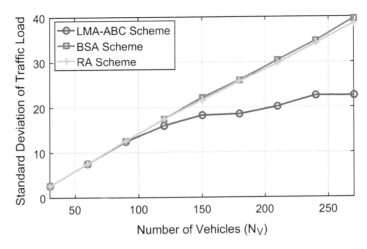

Fig. 5.2 Traffic load balancing performances of schemes with different association methods

a more balanced traffic load distribution. As shown in Fig. 5.2, with more vehicles in the scenario, the traffic load STD increases due to the uneven distribution of vehicles. However, the proposed *LMA-ABC* scheme can always achieve a better load balancing performance when compared to the BSA and RA schemes, and the performance gap enlarges with increasing N_V. For the BSA scheme, the network load unbalancing problem exacerbates with more vehicles in the scenario, especially in some intersection areas where a large number of vehicles are associated with the same AP. The traffic load STD of the RA scheme is slightly smaller than that of the BSA scheme, but the difference is negligible. On the other hand, the *LMA-ABC* scheme achieves a significantly smaller traffic load STD comparing to the BSA and RA schemes since it considers the potential future gain to alleviate network load imbalance.

Figure 5.3 shows the delay performance of the schemes with different association methods. We can observe that for all the schemes, the delivery delay increases with the number of vehicles, which is reasonable since fewer resources are available for each content delivery. With a small N_V, vehicles are sparsely distributed and network resources are sufficient for content deliveries, and thus, all the schemes can achieve a good delay performance. In overloaded networks with a large N_V, the delay performance for BSA and RA schemes degrades significantly. This is caused by the traffic load unbalancing problem as explained in Fig. 5.2. On the other hand, the *LMA-ABC* scheme can outperform the other two schemes and achieve the lowest delay since it considers channel quality, load balancing, and vehicle mobility to guarantee balanced user association with good content delivery performance.

Figure 5.4 shows the delay performance of the schemes with different bandwidth allocation methods. When N_V increases, the delivery delay of the *LMA-ABC* scheme increases, which is consistent with the results in Fig. 5.3. However, the delivery delay of the EBA and ETBA schemes shows a decreasing trend. For

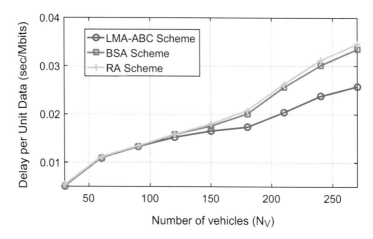

Fig. 5.3 Content delivery delay performance of schemes with different association methods

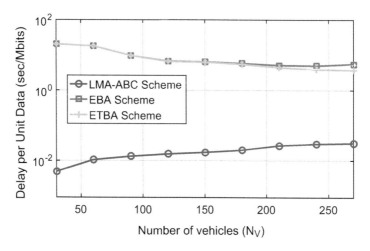

Fig. 5.4 Content delivery delay performance of schemes with different bandwidth allocation methods

the EBA and ETBA schemes, the traffic load and the diminishing gain effect are not taken into account for bandwidth allocation. Therefore, some vehicles with unsatisfactory channel conditions or a large content size are not allocated sufficient bandwidth resources to support the content delivery. In light-loaded networks, the undesirable bandwidth allocation has a non-negligible impact on the overall delay performance. With a large N_V, although the overall delivery delay increases, the average delay per unit data decreases since the impact of the undesirable bandwidth allocation diminishes. Furthermore, the *LMA-ABC* scheme significantly outperforms the other two schemes since it considers the diminishing gain effect and load balancing to guarantee superior delay performance.

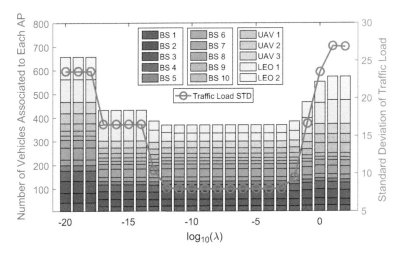

Fig. 5.5 Traffic load balancing performance of the *LMA-ABC* scheme with different λ

To investigate the impact of λ on the performance of the *LMA-ABC* scheme, we first demonstrate the traffic load balancing performance with different values of λ. Focusing on the scenario with 300 vehicles, the number of vehicles associated with each AP and the traffic load STD is plotted in Fig. 5.5. From the figure, we can observe that:

- When λ is small, the impact of the future potential gain is negligible when calculating the delay performance gain of each sub-channel allocation. Therefore, all the vehicles prefer to associate with the APs with the best SNR, leading to a poor load balancing performance with a large traffic load STD. Furthermore, due to the diminishing gain effect, each AP prefers to serve multiple vehicles instead of allocating all its resources to the same vehicle. Therefore, in this case, the total number of vehicle-to-AP associations is large.
- When λ increases, the impact of the future potential gain is non-negligible. As shown in the figure, the traffic balancing performance is significantly enhanced with decreasing STD. In this case, vehicles are generally associated with less crowded APs with good channel conditions, and the performance gain is limited for a vehicle to associate with an already congested AP. For instance, for a vehicle connecting to a BS with a good content delivery performance, it does not need the cooperation from a congested UAV or LEO satellite and thus leading to a decreased total number of vehicle-to-AP associations.
- When the value of λ keeps increasing to a sufficiently large number, the importance of the future potential gain is overemphasized. Therefore, vehicles always prefer less crowded APs regardless of the channel conditions. In this case, some vehicles are connected to APs with a poor achievable SNR and require more cooperative APs to enhance the content delivery delay performance, leading to an increasing number of vehicle-to-AP associations. Due to the

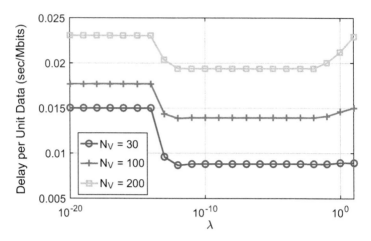

Fig. 5.6 Impact of λ on the delay performance of the *LMA-ABC* scheme

different coverage radii of the APs (e.g., LEO satellites' coverage is much larger than that of an LTE BS), the number of vehicles associated with each AP varies significantly with a large traffic load STD.

Figure 5.6 shows the impact of λ on the achievable delay performance of the *LMA-ABC* scheme. When λ increases, the delay performance first decreases due to the benefit from load balancing. When λ keeps increasing, the load balancing is overemphasized and the impact of channel conditions is underestimated, leading to a delay performance degradation, which is consistent with the results in Fig. 5.5. We can also see from the figure that when λ becomes too large, the value of λ has a larger impact on the delay performance for the case with more vehicles, since the network resources are insufficient and the improper allocation can greatly affect the overall delay performance. Therefore, the value of λ should be carefully optimized to achieve a good trade-off between the load balancing and channel conditions.

In Fig. 5.7, we compare the delay and simulation time performances of the *LMA-ABC* scheme with different bandwidth allocation granularities. As mentioned in Sect. 5.3, the bandwidth allocation is performed in the units of sub-channels. Therefore, the bandwidth allocation granularity greatly determines the delay performance and the decision-making complexity. Generally, with coarser bandwidth allocation granularities, e.g., allocating 5 or 10 sub-channels at a time, the delay performance degrades as shown in Fig. 5.7a, but the corresponding computational complexity also decreases as can be seen in Fig. 5.7b. Therefore, when the number of vehicles is small and the network resources are sufficient to support the vehicular content requests, we can moderately reduce the allocation granularities to achieve a good delay performance with a low complexity. For example, as shown in Fig. 5.7a, when the vehicle number is below 90, allocating 5 sub-channels at a time can achieve the same delay performance as the case which allocates one sub-channel at a time, but with a much lower delay. Therefore, the bandwidth allocation granularity should

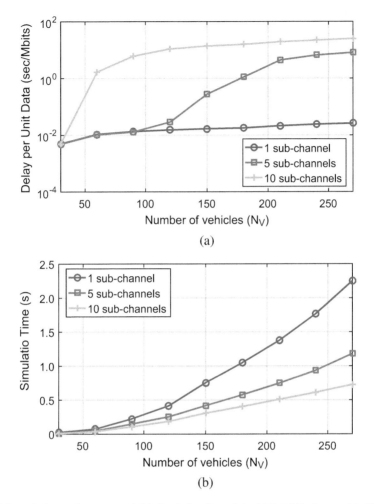

Fig. 5.7 (**a**) Delay performance and (**b**) simulation time of the *LMA-ABC* scheme with different bandwidth allocation granularities

be carefully designed in different scenarios to balance between the requirements on delay performance and time complexity.

5.5 Summary

In this chapter, we have investigated the cooperative content delivery in the SAGVN to minimize the overall content delivery delay. As the formulated *ABC* problem is intractable due to the tightly coupled continuous and integer variables, we have proposed an *LMA-ABC* scheme to jointly optimize the vehicle-to-AP association, bandwidth allocation, and content delivery ratio. With the analysis on the content delivery ratio optimization and the diminishing gain effect for bandwidth allocation, the *LMA-ABC* scheme can effectively solve the *ABC* problem by considering user fairness, load balancing, and vehicle mobility. Simulation results demonstrate that the proposed *LMA-ABC* scheme can significantly reduce the cooperative content delivery delay and achieve load balancing compared to the benchmark schemes.

References

1. H. Wu, J. Chen, C. Zhou, W. Shi, N. Cheng, W. Xu, W. Zhuang, X. Shen, Resource management in space-air-ground integrated vehicular networks: SDN control and AI algorithm design. IEEE Wirel. Commun. **27**(6), 52–60 (2020)
2. J. Liu, Y. Shi, Z.M. Fadlullah, N. Kato, Space-air-ground integrated network: a survey. IEEE Commun. Surv. Tutorials **20**(4), 2714–2741 (2018)
3. F. Lyu, P. Yang, H. Wu, C. Zhou, J. Ren, Y. Zhang, X. Shen, Service-oriented dynamic resource slicing and optimization for space-air-ground integrated vehicular networks. IEEE Trans. Intell. Transp. Syst. (2021). https://doi.org/10.1109/TITS.2021.3070542
4. Q. Ye, B. Rong, Y. Chen, M. Al-Shalash, C. Caramanis, J.G. Andrews, User association for load balancing in heterogeneous cellular networks. IEEE Trans. Wirel. Commun. **12**(6), 2706–2716 (2013)
5. Q. Han, B. Yang, G. Miao, C. Chen, X. Wang, X. Guan, Backhaul-aware user association and resource allocation for energy-constrained HetNets. IEEE Trans. Veh. Technol. **66**(1), 580–593 (2016)
6. A. Khalili, S. Akhlaghi, H. Tabassum, D.W.K. Ng, Joint user association and resource allocation in the uplink of heterogeneous networks. IEEE Wirel. Commun. Lett. **9**(6), 804–808 (2020)
7. W. Wu, N. Chen, C. Zhou, M. Li, X. Shen, W. Zhuang, X. Li, Dynamic RAN slicing for service-oriented vehicular networks via constrained learning. IEEE J. Sel. Areas Commun. **39**(7), 2076–2089 (2021)
8. Z. Li, C. Wang, C.-J. Jiang, User association for load balancing in vehicular networks: an online reinforcement learning approach. IEEE Trans. Intell. Transp. Syst. **18**(8), 2217–2228 (2017)
9. C. Chaieb, Z. Mlika, F. Abdelkefi, W. Ajib, On the optimization of user association and resource allocation in HetNets with mm-wave base stations. IEEE Syst. J. **14**(3), 3957–3967 (2020)
10. A. Wolf, P. Schulz, M. Dörpinghaus, J.C.S. Santos Filho, G. Fettweis, How reliable and capable is multi-connectivity? IEEE Trans. Commun. **67**(2), 1506–1520 (2019)
11. H. Wu, F. Lyu, C. Zhou, J. Chen, L. Wang, X. Shen, Optimal UAV caching and trajectory in aerial-assisted vehicular networks: a learning-based approach. IEEE J. Sel. Areas Commun. **38**(12), 2783–2797 (2020)
12. A. Al-Hilo, M. Samir, C. Assi, S. Sharafeddine, D. Ebrahimi, UAV-assisted content delivery in intelligent transportation systems-joint trajectory planning and cache management. IEEE Trans. Intell. Transp. Syst. **22**(8), 5155–5167 (2021)
13. H. Wu, J. Chen, W. Xu, N. Cheng, W. Shi, L. Wang, X. Shen, Delay-minimized edge caching in heterogeneous vehicular networks: a matching-based approach. IEEE Trans. Wirel. Commun. **19**(10), 6409–6424 (2020)
14. J. Chen, H. Wu, P. Yang, F. Lyu, X. Shen, Cooperative edge caching with location-based and popular contents for vehicular networks. IEEE Trans. Veh. Technol. **69**(9), 10291–10305 (2020)

15. M. Chen, M. Mozaffari, W. Saad, C. Yin, M. Debbah, C.S. Hong, Caching in the sky: proactive deployment of cache-enabled unmanned aerial vehicles for optimized quality-of-experience. IEEE J. Sel. Areas Commun. **35**(5), 1046–1061 (2017)
16. J. Du, C. Jiang, J. Wang, Y. Ren, S. Yu, Z. Han, Resource allocation in space multiaccess systems. IEEE Trans. Aerosp. Electron. Syst. **53**(2), 598–618 (2017)
17. Assembly, ITU Radiocommunication, Satellite system characteristics to be considered in frequency sharing analyses within the fixed-satellite service, in *ITU-R S.1328*, Sept 2002

Chapter 6
Conclusions and Future Research Directions

Abstract In this chapter, we conclude the main results and contributions of this monograph and present some future potential research directions.

6.1 Conclusions

In this monograph, we have investigated mobile edge caching-assisted vehicular content delivery in HetVNets. Specifically, three content caching and delivery schemes have been proposed for different HetVNet scenarios, i.e., the many-to-one matching-based content placement scheme in the terrestrial HetVNet, the *LB-JCTO* scheme, which jointly optimizes UAV content caching, UAV content delivery, and UAV trajectory design in the AGVN, and the *LMA-ABC* scheme, which jointly addresses the user association, bandwidth allocation, and content delivery ratio optimization in the SAGVN. Heterogeneous network characteristics, UAV energy consumption, vehicle mobility patterns, and content file properties are considered in the proposed schemes. The main contributions of this monograph are summarized as follows:

1. The general framework of mobile edge caching-assisted HetVNets has been proposed, where heterogeneous network segments can cooperate to enhance the vehicular content delivery performance. Specifically, the impact of factors including content popularity, vehicle mobility, network service disruptions, and APs' caching capacity constraints on the achievable content delivery delay performance has been theoretically analyzed. To resist the impact of intermittent network connections, content coding is leveraged with optimized coding parameters to encode content files into packets, and the PRAI transmission mode is applied to strike a good balance between the delay performance and the offloading ratio. Based on the analysis, a matching-based scheme with multi-objective two-sided preference lists has been proposed to optimize the content placement in heterogeneous APs. This provides a theoretical basis for future studies related to edge caching in heterogeneous networks such as the SAGVN.

H. Wu et al., *Mobile Edge Caching in Heterogeneous Vehicular Networks*,
SpringerBriefs in Computer Science, https://doi.org/10.1007/978-3-030-88878-7_6

2. We have investigated the joint optimization of content caching, content delivery, and UAV trajectory design in the AGVN. To find the optimal solution in real time to maximize the overall network throughput under the UAVs' energy constraints, we have proposed an *LB-JCTO* scheme. *LB-JCTO* is an offline optimization and learning for online decision framework, in which a CNN-based learning model is trained to facilitate online decisions under the supervision of offline optimized targets obtained by the CBTL algorithm. The problem formulation of JCTO and the optimization process of CBTL can provide a theoretical basis for future studies related to mobile edge caching-enabled UAV systems. In addition, we believe the principle of offline optimization and learning for online decisions can also be valuable for other complicated resource management in future heterogeneous networks.

3. We have proposed an *LMA-ABC* scheme for effective cooperative content delivery in the SAGVN, which jointly optimizes the user association, spectrum resource allocation, and content delivery ratio. Specifically, the *LMA-ABC* scheme aims to reduce the overall content delivery delay by taking user fairness, load balancing, and vehicle mobility into account. With the proposed scheme, heterogeneous network resources in the SAGVN can be efficiently exploited to fully unleash their differential merits, and the overall content delivery delay can be significantly reduced, which is of capital importance for CAV services. Besides, the problem formulation of *ABC* and the optimization process of the *LMA-ABC* can provide a theoretical basis for future research on multi-connectivity-based SAGVNs.

6.2   Future Research Directions

Toward enhancing the service quality of vehicular applications in the mobile edge caching-assisted SAGVN, there still exist many open research issues. For the future research, there are some interesting and promising research directions listed as follows:

1. **Service-oriented multi-dimensional resource orchestration:** In mobile edge caching systems, the emergence of advanced vehicular applications requires the joint management of multi-dimensional resources (i.e., caching, computing, and communication resources) to further improve vehicular service quality. For instance, for video-oriented mobile applications, caching resources are required for storing the video content at the edge networks, computing resources are required to compress the cached video content with different quality levels, and communication resources are utilized for the video content delivery to end users. Basically, different vehicular services have differentiated requirements on the multi-dimensional resources, each of which can be provided by different SAGVN network infrastructure (e.g., terrestrial BSs, UAVs, satellites). To achieve the optimal network performance and increase resource utilization efficiency, a

service-oriented resource orchestration scheme should be developed in the mobile edge caching-assisted SAGVN by considering the following factors (1) the space, air, and terrestrial network segments have distinctive characteristics in terms of communication delay, throughput, coverage, computing capability, jitter, and so on, (2) multiple types of mobility introduced by the vehicles, UAVs, and LEO satellites lead to dynamic network resource availability, and the spatial–temporal variations in communication link conditions affect the required resources to satisfy the QoS requirements, and (3) due to the differentiated QoS requirements, the priorities for various vehicular applications to access the network resources from different network segments are different. Therefore, service-oriented multi-dimensional resource orchestration solutions are imperative but challenging.

2. **SDN-based control architecture in the SAGVN:** To efficiently manage the resources from space, air, and ground network segments in mobile edge caching systems, a flexible, reliable, and scalable control architecture is required. Recently, the SDN-based hybrid and hierarchical control architecture has attracted increasing research interests. In this control architecture, the placement of SDN controllers is critical due to its significant impact on network performance such as caching resource utilization efficiency, communication latency, load balance of controllers, network availability, and energy consumption. The SDN controller placement problem (SCPP) involves the determination of the optimal number of SDN controllers, the best placement locations, and the division of the control domains. Basically, the SCPP optimization should consider factors such as capacity of controllers, the network traffic loads, and control latency requirements. Therefore, the problem of how to strategically place the SDN controllers in the mobile edge caching-assisted SAGVN should be carefully investigated to guarantee reliable network control with low signaling overhead and short control latency.

3. **AI-based network management:** In the SAGVN, modeling the dynamic and complex network system is painstaking, if not impossible. Furthermore, traditional model- and optimization-based approaches are inadequate for the multi-dimensional resource management in mobile edge caching systems due to the high complexity. Innovative AI-based engineering solutions are necessary to make high-quality and real-time decisions to keep pace with the dynamic environment. Particularly, distributed AI can be adopted to enable intelligent decision-making at different granular levels through parallel training processes. When designing the distributed AI-based network management schemes, the appropriate splitting of data and model, communication overhead of model updating, and the caching/computing capabilities of different network nodes should be taken into consideration. Some recently proposed distributed AI paradigms, e.g., federated learning and split learning, can also be considered.

Index